'A wonderfully absorbing book about an engineer and her passion for construction.'

Professor Mark Miodownik, materials engineer, broadcaster and author of *Stuff Matters*

'Roma makes the complex principles of structural engineering accessible to everyone with clear explanations and engaging illustrations. It has made me suddenly look at every building I pass in a new way.'

Ellen Stofan, former Chief Scientist at NASA

'A charming tribute to one woman's love of engineering. Full of fascinating facts and personal stories.'

Angela Saini, author of *Geek Nation* and *Inferior: How Science Got Women Wrong – and the New Research That's Rewriting the Story*

'A necessary reminder of the wonderful human ingenuity behind the world's greatest engineering projects, from Roman aqueducts to London's magnificent Shard. *Built* will inspire readers of every stripe.'

Erica Wagner, author of *Chief Engineer*

'A book about real engineering written by a real engineer who can really write.'

Henry Petroski, author of *The Road Taken*

'A passionate, often personal love letter to the science of structure. Whether exploring Pompeii or assembling pineapple upside-down

cake, Agrawal's journey will change the way you look at the structures you take for granted.'

Jennifer Ouellette, author of *The Calculus Diaries*

'Roma Agrawal is a much-needed role model for the next generation of engineers. Most girls never give a thought to civil engineering as a career. This book will change that.'

Rob Eastaway, mathematician and author of
Any Ideas? Tips and Techniques for Creative Problem Solving

'Roma Agrawal makes engineering thrilling, accessible and thoroughly enjoyable. *Built* is another great example of how women can write brilliantly about so-called male subjects. Bring on the female engineers!'

Andrea Wulf, historian and author of *Chasing Venus:
The Race to Measure the Heavens*

'How come we have skyscrapers, bricks, bridges and tunnels? It's down to structural engineers, but who are these awesome makers? Men in hard hats? Not necessarily! Roma Agrawal has helped to design some iconic structures that grace our cities. Here she gives an insider's captivating account, so rich in history and so full of verve that she is bound to make others want to follow in her footsteps. *Built* reveals how human ingenuity keeps us safe from the raw forces of nature, all in a style that is down-to-earth as well as funny and personal.'

Uta Frith, psychologist and author of *Autism:
Explaining the Enigma*

'Hurrah! A joyful book that dissects our architecture to reveal the decisions behind the anatomy we rely on every day.'

Gaia Vince, journalist, broadcaster and author
of *Adventures in the Anthropocene*

BUILT

BUILT

*The hidden stories
behind our structures*

Roma Agrawal

BLOOMSBURY PUBLISHING
LONDON · OXFORD · NEW YORK · NEW DELHI · SYDNEY

BLOOMSBURY PUBLISHING
Bloomsbury Publishing Plc
50 Bedford Square, London, WC1B 3DP, UK

BLOOMSBURY, BLOOMSBURY PUBLISHING and the Diana logo are
trademarks of Bloomsbury Publishing Plc

First published in Great Britain in 2018
This edition published in 2019

Text and illustration copyright © Roma The Engineer Ltd, 2018
For picture credits, see page 283

Roma The Engineer Ltd has asserted its right under the Copyright,
Designs and Patents Act, 1988, to be identified as Author of this work

British Library Cataloguing-in-Publication Data
A catalogue record for this book is available from the British Library

ISBN: 978-1-4088-7037-2

2 4 6 8 10 9 7 5 3 1

Typeset by Newgen Knowledge Works Pvt. Ltd., Chennai, India
Printed and bound in Great Britain by CPI Group (UK) Ltd, Croydon CR0 4YY

To find out more about our authors and books visit www.bloomsbury.com
and sign up for our newsletters

For Maa,
and for little Samuel.

CONTENTS

STOREY

With one hand, I clutched my precious stuffed-toy cat, afraid that I would lose it. With the other, I clung to my mother's skirt. Terrified and exhilarated by the new, strange, unknown world in constant motion around me, I held on to the only two things that felt familiar.

When I think of Manhattan now, I am always taken back to my first visit, as an impressionable toddler: the funny smell of the car exhausts, the shouts of the streetside lemonade vendors, the swarm of people rushing by, bumping into me unapologetically. It was an overwhelming experience for a child who lived far from the big city. Here, instead of open sky, I saw towers of glass and steel blocking out the sun. What were these monstrous things? How could I climb them? What did they look like from above? I turned my head left and right as my mother dragged me along the busy streets. Stumbling after her with my head raised, I was transfixed by these pillars that reached towards the clouds.

At home, with my miniature cranes, I stacked building blocks to recreate what I had seen. At school, I painted tall rectangles on big sheets of paper in bright, bold colours. New York became part of my mental landscape as I visited and revisited the place over the years, admiring new towers that appeared on the ever-changing skyline.

For a few years we lived in America, while my father worked as an electrical engineer. We didn't live in one of the soaring skyscrapers that so impressed me on my visits to Manhattan, however, but in a creaking wooden house among the hills upstate. When I was six, my father gave up engineering to look after the family business in Mumbai, and I went to live in a seven-storey concrete tower that looked out towards the Arabian Sea. When my Barbie dolls finally arrived safe and sound at my new home, after a long sea journey in a storage container, it was of course essential that they were made comfortable. Pop helped me reassemble my cranes, laying out a large white sheet so I wouldn't lose any pieces. Making loud, whirring noises, I lifted long plastic tubes and manoeuvred pieces of card into position, building a house for my dolls. My first step, perhaps, towards a career in engineering.

Having an American accent and – as you'll soon discover if you haven't already – a tendency to be a bit geeky, I found my new school a challenge at first. I was teased by some for being a 'scholar'. But gradually I found friends and teachers that 'got' me. Through large gold-framed glasses, I eagerly read physics, maths and geography textbooks, and I loved art class, although I struggled with chemistry, history and languages. Mom, who had studied maths and science at university and had worked as a computer programmer, encouraged my growing interest in science and maths, assigning me extra homework and reading. Throughout my school years I loved these subjects best and I resolved to be an astronaut or an architect when I grew up.

Back then, I'd never even heard the term 'structural engineer', and never imagined that one day I would play a part in designing a magnificent skyscraper – The Shard.

Since I loved learning so much, my family decided I should finish my schooling in another country, as it would be a great opportunity to broaden my horizons. And so, aged fifteen, I moved to London to study maths, physics and design at A level. Another new school in a new country, but this time I quickly sought out kindred spirits – girls who found Faraday's law as fascinating as I did, and who experimented in the lab just for fun. Brilliant teachers paved my way to studying physics at university, and I moved to Oxford.

At school, physics made sense to me. At university, it didn't – at least to begin with. Light was both a wave and a collection of particles? Space-time could be curved?? Time travel was mathematically possible?! I was hooked, but it was tough stuff to get my head around. Academically, I always felt like I was a few steps behind my peers. It was a real reward when I finally figured out how something worked. I balanced hours in the library with ballroom and Latin dance lessons, learning to wash clothes and to cook (though perhaps not all that skilfully, as you'll see), and generally fending for myself. I was enjoying physics; my childhood dreams of going into space or becoming an architect became distant memories. At the same time, however, I had little idea of what I wanted to do with my life.

Then, one summer, I worked in the physics department at the University of Oxford, drawing up plans of all the fire-safety features in the various buildings. The task in itself was hardly world-changing, but the people who sat around me were working on projects that were. They were engineers, and their job was to design the equipment that physicists could use to seek out the particles that define how our world works. As you might imagine, I badgered them with questions and was astonished at what their jobs entailed. One was designing a metal holder for a glass lens – a simple task, you might think, except that the whole apparatus had to be cooled to -70° Celsius. Metal contracts more than glass, and unless the holder was cleverly and carefully designed, the cooling metal would crush the lens. It was just a tiny piece in a immense maze of machinery, but a complex and creative challenge. I spent hours of my free time trying to figure out how I might solve the problem.

Suddenly, it became very clear to me: I wanted to use physics and maths to solve practical problems and, in the process, help the world in some way. And it was at this point that my childhood love of skyscrapers re-emerged from the depths of my memory. I would be a structural engineer and design buildings. To make the transition from physicist to engineer, I studied at Imperial College London for a year, graduated, got a job – and began my life as an engineer.

As a structural engineer, I am responsible for making sure that the structures I design stand up. In the past decade I have worked on an amazing variety of constructions. I was part of the team that designed The Shard – the tallest tower in Western Europe – spending six years working out the sums for its open-air spire and foundations; I worked on a fancy footbridge in Newcastle, and the curving canopy at Crystal Palace station in London. I've designed hundreds of new apartments, brought a Georgian townhouse back to its former glory, and ensured an artist's sculpture was stable. Whilst my job involves using maths and physics to create things (which in itself is incredible fun), it is also so much more. For a start, a modern engineering project is an enormous piece of teamwork. In the past, engineers like Vitruvius (who wrote the first treatise on architecture) or Brunelleschi (who built the breathtaking dome that crowns Florence's cathedral) were known as master builders. They knew about every discipline necessary for construction. Nowadays structures are more complex and technically advanced, and no single person can design every aspect of a project. Each of us has an area of specialisation, and the challenge is to bring everybody together in an intricate and quietly frenetic dance that weaves together materials, physical effort and mathematical calculations. With the architects and other engineers, I brainstorm design problems. Our drawings assist site managers, and surveyors calculate costs and consider logistics. Workers on site receive materials and reshape them to create our vision. At times, it's hard to imagine that all this sometimes chaotic activity will resolve into a solid structure that will last for decades, or even centuries.

For me, each new structure I design becomes personal, as 'my' building grows and takes on its own individual character. At first we communicate through a few rough drawings, but gradually I discover what will prop it up, and how it will stand tall and be able to evolve with the changing times. The more time I spend with it, the more I come to respect, even love it. Once complete, I get to meet her in person, and walk around her. Even after that, as far as I'm concerned, we have an ongoing commitment to one another, and I watch from afar as other people take my place and develop their own relationships with my creation, making the building their home or workplace, protected from the outside world.

Of course, my feelings for the structures I have worked on are particularly personal, but in fact all of us are intimately connected to the engineering that surrounds us – the streets we walk on, the tunnels we rush through, the bridges we cross. We use them to make our lives easier, and we look after them. In return, they become a silent but crucial part of our existence. We feel charged and professional when we walk into a glass skyscraper with neat rows of desks. The speed of our journey is emphasised by steel rings flying past the windows of an underground train. Uneven brick walls and cobbled stone pathways remind us of the past, of the history that has gone before us. Structures shape and sustain our lives and provide the canvas of our existence. We often ignore or are unaware of them, but structures have stories. The tense cables stretching above a massive bridge across a river; the steel skeleton beneath the glass skin of a tall tower; the conduits and tunnels burrowing beneath our feet – these things make up our built world, and they reveal a lot about human ingenuity, as well as our interactions with each other, and with Nature. Our ever-changing, engineered universe is a narrative full of stories and secrets that, if you have the ears to listen, and the eyes to see, is fascinating to experience.

My hope is that, through this book, you too will discover these stories and learn these secrets. That a new understanding of our surroundings will change the way you look at the hundreds of structures you move over, below and through every day. That you will see your home, your city, town or village, and the countryside beyond with a new sense of wonder. That you will see your world through different eyes – the eyes of an engineer.

FORCE

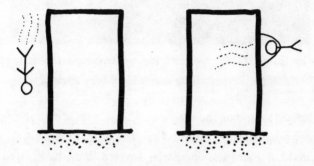

It's a peculiar feeling when you step onto or into something you've designed. My first project after leaving university was the Northumbria University Footbridge in Newcastle, England. For two years, I worked with the architects' plans, helping to make their vision a reality, covering hundreds of pages with calculations and creating countless computer models. Eventually it was constructed. Once the cranes and diggers had moved on, I finally had the opportunity to stand on the steel structure I had helped to create.

The Northumbria University Footbridge, built in 2007 to link the two main sites of the university in Newcastle Upon Tyne, England.

Briefly, I stood on the solid ground just in front of the bridge, before taking a step forward. I remember that moment: I was excited but I also felt disbelief – amazed that I had played a role in making this beautiful bridge stand, so that hundreds of people could walk across it every day. I looked up at its tall steel mast and the cables radiating from it, supporting the slim deck safely above the motorway traffic – it held its own weight, and mine, effortlessly. Balustrades, carefully angled to make them difficult to climb, reflected the cold sunlight. Below me, cars and trucks whizzed past, oblivious to this young engineer standing proudly on 'her' bridge, marvelling at her first physical contribution to the world.

It was, of course, steadfast beneath my feet. After all, those numbers and models I had carefully executed to calculate the forces my bridge would be subjected to had been checked and

Standing upon the Northumbria University Footbridge, my first project as an engineer.

re-checked. Because, as engineers, we can't afford to make mistakes. I'm conscious that every day thousands of people will use structures that I have designed: they will cross them, work in them, live in them, oblivious to any concern that my creations could let them down. We put our faith and our feet (often quite literally) on engineering, and it is the engineer's responsibility to render things robust and reliable. For all that, history has shown us that things can go wrong. On the afternoon of 29 August 1907 residents of Quebec City thought they had just been shaken by an earthquake. In fact, 15 km away, something far more unthinkable was happening. On the banks of the Saint Lawrence River the sound of ripping metal tore through the air. The rivets that held together a bridge under construction had snapped, catapulting over the heads of terrified workers. The steel supports for the structure folded as if they were made from paper, and the bridge – with most of its builders – plunged into the river. One of the worst bridge collapses in history, it is a brutal example of how mismanagement and miscalculation can end in disaster.

*

Bridges expand cities, bring people together and promote commerce and communication. The idea of building a bridge across the Saint Lawrence had been debated in parliament since the 1850s. The technical challenge, though, was huge: the river was three kilometres wide at its narrowest point, with deep, fast-moving water. In winter the water froze, creating piles of ice as high as 15 m in the river channel. Nonetheless, the Quebec Bridge Company was eventually set up to undertake the project, and work on the foundations began in 1900.

The company's chief engineer, Edward Hoare, had never before worked on a bridge longer than 90 m (even the original plans for this project called for a 'clear span length' – i.e. a length of bridge without any supports – of just over 480 m). So the fateful decision was made to enlist the services of Theodore Cooper as consultant. Cooper was widely regarded as one of the best bridge builders in America, and had written an award-winning paper on the use of steel in railway bridges. Theoretically he must have seemed like the ideal candidate. But there were problems from the start. Cooper lived far away in New York, and his ill health meant he rarely visited the site. Yet he insisted on being personally responsible for inspecting the steel fabrication and construction. He refused to have his design checked by anyone and relied on his relatively inexperienced inspector, Norman McLure, to keep him informed of progress on site. Construction on the steel structure began in 1905, but over the next two years McLure became increasingly worried about how the build was progressing. For a start, the pieces of steel arriving from the factory were heavier than he expected. Some of them were even bowed rather than straight because they were buckling under their own weight. Even more worryingly, many of the steel pieces installed by the workers had already deformed even before the bridge was complete, a sign that they were not strong enough to carry the forces flowing through them.

This deformation was the result of Cooper's decision to change the design of the bridge away from its original plans, increasing the length of the central span (the unsupported middle of the bridge) to nearly 549 m. Ambition may have clouded Cooper's judgement: in making the decision he might have hoped to become the engineer responsible for the longest-spanning cantilever bridge in the world, an honour held at the time by the Forth Bridge in Scotland. The larger the span of a bridge, the more material you need to build it, and the heavier it becomes. Cooper's new design was about 18 per cent heavier than the original, yet without paying enough attention to the calculations, he decided that the structure was still strong enough to carry this extra weight. McLure disagreed, and the two men argued about it in an exchange of letters. But nothing was resolved.

Finally, McLure became so concerned that he suspended construction and set off by train to New York to confront Cooper. In his absence, an engineer on site overturned his instructions and the workforce went back to assembling the bridge, with tragic results. In just fifteen seconds, the entire south half of the bridge – 19,000 tonnes of steel – collapsed into the river, killing 75 of the 86 people working on the structure.

The scene of devastation following the collapse during construction of the Quebec Bridge in 1907, spanning the Saint Lawrence River, Quebec City, Canada.

Many problems and mistakes contributed to the bridge's collapse. In particular, the disaster revealed the dangers of putting huge power in the hands of one engineer without supervision. In Canada and elsewhere, organisations of professional engineers were set up to regulate the profession and try and prevent a repeat of the Quebec Bridge mistakes. Ultimately, however, much of the responsibility lies with Theodore Cooper, who underestimated the weight of the bridge. In the end, the way it was engineered meant it was just too feeble to hold itself up.

*

The abrupt devastation of the Quebec Bridge demonstrates the catastrophic effect gravity can have on a faulty human construction. A major part of the engineer's job is figuring out how structures can withstand the manifold forces determined to push, pull, shake, twist, squash, bend, rend, split, snap or tear them apart. Grappling with gravity is therefore a key consideration on many projects. It is the omnipresent force that holds the solar system together, and which attracts everything on our planet towards its centre. This creates a force within every object, which we call its weight. This force *flows* through the object. Think about the weight of different parts of your body. The weight of your hand acts on your arm, pulling on your shoulder then pushing into your spine. Flowing down the spine, the force reaches your hips, and here, at the pelvic bone, the force splits into two, flowing into each of your legs and down into the ground. In much the same way, if you build a tower from straws and pour water on top of it, the water will stream through the different pathways it finds, dividing where more than one option is available.

When planning a structure, then, it is vital for an engineer to understand where the force is flowing, what kind of force it is, and then make sure that the structure transmitting the force is strong enough for the job.

There are two main types of forces that gravity (and also other phenomena such as wind and earthquakes) creates in structures: *compression* and *tension*. If you roll a piece of thick paper into a cylindrical tube, stand it vertically on a table, and then put a book on top of it, the book pushes down on the tube. The *force* with which it does this (which is its mass multiplied by gravity) flows through the tube down to the table – just as your weight flows through your leg. The tube (like your leg) is in *compression*.

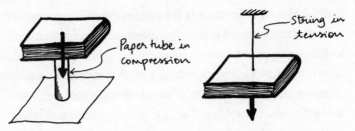

Supporting a book using compression (above left) and tension (above right).

Conversely, if you take a piece of string, tie the book to one end and hold the other, the suspended book – still experiencing the force of gravity – is now pulling on the string. The force of the book flows up into the string, which is said to be in *tension*. This is the same effect that the weight of your hand has on your arm.

In the first example, the book doesn't crash down onto the table because the paper tube is strong enough to resist the *compression* it feels. In the second example, it remains safely suspended because the piece of string is strong enough to resist the *tension* it feels.

To cause a collapse, use a heavier book. The new force exerted by this book on its support is larger because the weight of the book has increased. The tube is no longer strong enough, so it crushes and the book falls to the table. Similarly, if you try suspending the heavier book, the tension is too big for the string. The string snaps and the book plummets.

The forces in a bridge arise from its own weight, and from the weight of the people and vehicles that travel over it. When working on the Northumbria University Footbridge, I did calculations to find out where the forces were in the structure. As a result, I knew exactly how much compression or tension was at work in each part of it. I used a computer model to test every section of my bridge, then calculated how big the steel needed to be so it didn't bend excessively, crush or snap.

*

The type of force and the way it flows depends on how the structure is assembled. There are two main ways this can be done. The first is known as the *load-bearing* system and the second as the *frame* system.

Our early ancestors' mud huts – which they made by forming mud into thick walls arranged in a circle or square – were built using the first method. The walls of these single-storey dwellings were solid, forming a load-bearing system: the weight of the structure was free to flow as compression throughout the mud walls. This is similar to the book resting on the paper tube, in which all sides of the tube are uniformly in compression. If additional storeys were added to the hut, at some point the compression

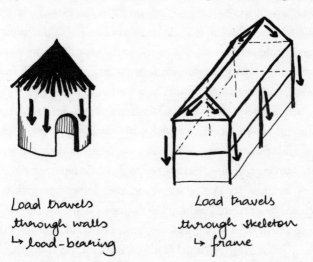

Load travels
through walls
↳ load-bearing

Load travels
through skeleton
↳ frame

Two ways to build a home, using load-bearing walls (above left) or a skeleton frame (above right).

would become too much for the load-bearing mud walls and they would crumble, just like the heavier book crushes the paper tube.

When our ancestors had access to trees, they built their homes using the frame system – by tying timber logs together to create a network or skeleton through which the forces are channelled. To protect the inside from the elements, animal skins or woven vegetation were suspended across the logs. Where mud huts had solid walls that bore the forces *and* protected the residents, the timber home had two distinct systems: the logs that carried the forces *plus* the 'walls' or the animal skins which carried no weight. The way in which forces are channelled is the fundamental difference between load-bearing and frame structures.

Over time, the materials we used to create load-bearing walls and frames for structures became more and more sophisticated. Load-bearing structures were made from brick and stone, which were stronger than mud. In the 1800s, after the Industrial Revolution, iron and steel could be manufactured at a large scale, and we started to use metals for building, rather than just for vessels and weaponry. Concrete was rediscovered (the Romans had known how to manufacture it, but that knowledge was subsequently lost when their empire fell). These moments of evolution changed our structures forever. Since steel and concrete are so much stronger than timber, and well-suited to creating large frames, we could build taller towers and longer bridges. Today, the largest and most complex structures – such as the graceful steel arch of Sydney Harbour Bridge, the triangular geometry of the Hearst Tower in Manhattan, and the iconic 'Bird's Nest' National Stadium built for the 2008 Beijing Olympics – are created using the frame system.

*

When I start designing a new building, I study the carefully crafted drawings from the architects which convey their vision of what the construction will look like once it's finished. Engineers soon develop a kind of X-ray vision, enabling them to see through the building in the picture to the skeleton it would need in order to resist gravity and the other forces that test it. I visualise where the building's spine will go, where the supporting bones need to be connected, and how big these need to be so the skeleton is stable. With a black marker pen, I sketch over the architects' drawings, adding bones to the flesh. The thick, black lines I add to the colourful drawings add a certain solidity. Inevitably, there is much discussion – sometimes quite spirited – between me and the architects: we need to compromise to find a solution. Often, I need a column where they have shown an open space; at other times they think I need structure where I don't – so I can give them more area. We have to understand each other's perspectives when technical problems arise: we must reach a balance between visual beauty and technical integrity. Eventually, we arrive at a design in which structure and aesthetic vision are (almost) in perfect harmony.

The frames in our structures are made up of a network of columns, beams and braces. Columns are the vertical sections of the skeleton; beams are the horizontal ones, and the pieces at other angles – the braces – are usually called 'struts'. If you look at a photograph of Sydney Harbour Bridge, for example, you'll see that it's formed of pieces of steel at all sorts of angles – a melee of columns, beams and struts. By understanding how columns and beams interact and support one another, how they attract forces and, most importantly, how they break, we can design them so they won't fail.

The Sydney Harbour Bridge, completed in 1930, built to carry rail, vehicular and pedestrian traffic between the North Shore and the central business district of Sydney, Australia.

Although columns have been used to resist gravity for millennia; the Greeks and Romans turned them into an art form. Much of the beauty and the solidity of the Parthenon in Athens comes from its outer row of fluted Doric marble columns. The remains of the Forum in Rome are dominated by monumental columns that support the fragile remnants of temples, or which simply strike upwards, stunted, towards the sky. Of course, the columns fulfilled a very important practical function – holding up structures – but this didn't stop ancient engineers from decorating them with carvings inspired by Nature and mythology. The Corinthian column, with its capital decorated by intricately curled leaves, was supposedly invented by the Greek sculptor Callimachus after he noticed an acanthus plant growing through and around a basket left upon the grave of a maiden of Corinth. There are dozens of examples of it dotted around the Forum, and it has remained a classic of civic architecture for centuries, grandly gracing the façade of the United States Supreme Court Building for example and, more humbly, the entrance to the Victorian block of flats where I live.

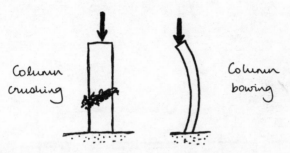

Two of the ways in which a column can fail, through crushing (above left) and bowing (above right).

Columns generally work by countering compression. One way they can fail is when they are squashed so much that the material they are made of simply gives up and crushes or crumbles – this is what happens to the paper tube when the heavier book is placed on top. The other way columns can fail is by bowing. Take a plastic ruler, stand it vertically on a table and then press down on it with the palm of your hand: you'll see it begin to bow. The more you push down, the further the ruler will bow – until finally it snaps.

There is a delicate balance to be struck when designing a column. You want it to be thin so that it doesn't take up too much space, but if it's too slender the load it carries can cause it to bow. At the same time, you want to use a material that's strong enough

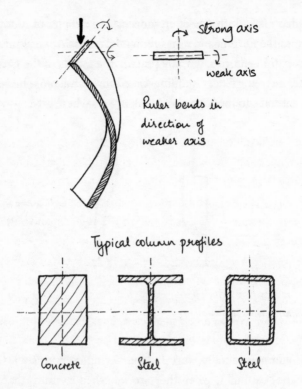

strong axis

weak axis

Ruler bends in direction of weaker axis

Typical column profiles

Concrete

Steel

Steel

Flexing a ruler shows how a slender structure bends along its weaker axis (top), whereas a column, whether it's made from concrete or steel, is shaped to resist bending in both axes (bottom).

to prevent crushing. The columns used in ancient structures tended to be stocky, chunky things most often made from stone, and were unlikely to fail by bowing. By contrast, our modern steel or concrete columns tend to be far more slender, making them mostly susceptible to bowing.

A ruler is wide in one direction and flat in the other: as you'll have seen as you pressed down, it bows about its much weaker axis. To stop this effect, modern steel columns are usually made in an H shape, and concrete columns in squares or rectangles so that both axes are comparably stiff – so the columns can resist larger loads.

*

Beams work differently. They form the skeleton of the floors. When we stand on a beam, it flexes slightly, channelling our weight across to the columns that support it. The columns in turn compress and transmit our weight to the ground. If you stand on the centre of a beam, half of your weight, and half the weight of the beam is transmitted to each end. The column then transmits that load downward. We don't want beams to bend too much when we stand on them, partly because it feels uncomfortable when the floor is moving below our feet – but also because they can fail. We need to make beams appropriately stiff; using depth, geometry or specific materials to strengthen them.

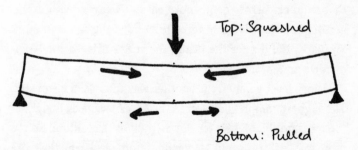

A beam flexes when it bears any weight, with the top of the beam being squashed and the bottom of the beam being pulled.

Typical beam profiles

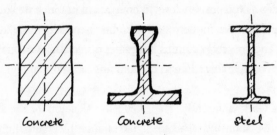

Concrete   Concrete   Steel

To resist this flexing, beams are made in specific shapes.

When a beam bends under a load, the load flows unevenly through it. The top portion of the beam is squashed, while the bottom portion is pulled: the top of the beam is in compression and the bottom is in tension. Try bending a carrot in your hands: as you curve it into a U-shape, the bottom eventually splits. This happens when the tension force in the bottom of the carrot is too big for the carrot to resist. If you repeat this with carrots of different diameters, you'll find that thinner carrots bend more easily. A carrot with a bigger diameter needs more force to bend it the same amount. Similarly, the deeper the beam, the stiffer it is, so the less it distorts under load.

Using clever geometry is another way to make a beam stiff. The highest compression force a beam experiences is right at the top, and the highest tension is right at the bottom. So the more material you put in the top or bottom of a beam, the more force it can take. By combining these two principles – depth and geometry – we arrive at the best shape for a beam: an I (i.e. in cross-section it resembles that letter), because the greatest amount of material is at the top and bottom, where the greatest forces flow. Most steel beams are I-shaped. (They are subtly different from H-shaped columns because they are deeper than they are wide, whereas H-shaped columns are closer to squares.) Concrete beams can also be made like this, but it is easier to pour concrete into a rectangular shape, so for reasons of cost and practicality most concrete beams are simple rectangles.

Large bridges like the Quebec Bridge are just too long to use a 'normal' I-shaped beam. To span the distance, such a beam would have to be so deep and heavy that it would be impossible to lift into place. Instead, we use another type of structure that harnesses the stability of triangles: the *truss*.

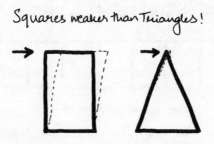

A square is an inherently weaker shape than a triangle.

Take four sticks and tape the corners together to make a square. Then push it sideways: the square becomes a diamond and collapses. Triangles, on the other hand, do not deform and collapse in the same way. A truss is a network of triangles made up of beams, columns and struts, which cleverly channels forces through its members. And in creating a truss we use smaller and lighter pieces with gaps in the middle, so we use less material than we would for an equivalent I-beam.

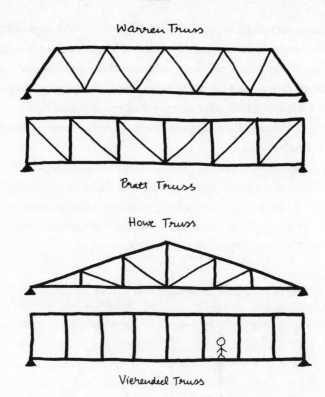

Warren Truss

Pratt Truss

Howe Truss

Vierendeel Truss

Most trusses are made up of smaller triangular shapes, although
occasionally some do use squares.

Trusses are easier to build because smaller pieces of steel can
be transported to the construction site and then joined together.
Most large bridges have trusses somewhere. Take a look at the
Golden Gate Bridge, for example: a pattern in the metal runs
along the sides at road level for the length of the bridge. It looks
like the letter N followed by a reversed N, one after another – a
careful arrangement of triangles forming a truss.

*

Gravity exerts a predictable pull on objects on the surface of the Earth. An engineer knows what it is, and can design columns, beams and trusses to resist it. But other, equally destructive forces are not so easily reduced to equations. One of these is the wind. Random, fluctuating, unpredictable, wind has challenged engineers throughout history, and it remains a problem all engineers have to solve if their structures are going to remain stable.

When I visited Athens, one of the monuments that I was most excited to see was a large white marble octagonal tower in the Roman Agora, just north of the Acropolis. Built around 50 BC by Andronicus of Cyrrhus, a Macedonian astronomer, the Horologion of Andronikos Kyrrhestes or 'Tower of the Winds' was a timepiece with eight sundials, a water clock and a wind vane. Taking a slow walk round the tower I could see that each of its sides had a relief at the top depicting one of eight Wind Gods, winged figures flying forwards with a stern or benign expression, and sometimes an amphora or garland of flowers in their arms. Originally a bronze statue of Triton stood on top of the tower and acted as a weathervane, pointing towards whichever Wind God was blowing.

The tower is a testament to the respect the Romans had for the wind and its potentially destructive force. The Roman master builder Marcus Vitruvius Pollio (born 80 BC), who is sometimes called 'the first architect', talks extensively about the importance of considering wind in *De Architectura*, his hugely influential ten-volume treatise on the design of structures. In Book 1 he tells us about the four main directions: *Solanus* (east), *Auster* (south), *Favonius* (west), *Septentrio* (north) – and the other four, which act in directions between the primary winds.

To me it seems amazing that Roman engineers already had such a deep understanding of how wind acts differently in different directions. Even though the way engineers calculate this is now much more sophisticated, the basis of our work was carved into the sculptures on that octagonal tower 2,000 years ago.

*

Wind acts on structures everywhere on our planet. When I am working on a construction that is less than 100 metres tall, I typically use a wind map. This is essentially a weather map with contours that tells me what the basic wind speed is at a particular location, created using data measured over decades. I take the basic wind speed and combine it with other numbers that define, say, how far the place is from the sea, how high up it is, and the variability of the surrounding terrain (how hilly it is or how many buildings there are). Formulae combine these factors to tell me how much wind a structure will feel in 12 different directions – every 30 degrees around a circle – which is not far off the eight directions enumerated by Vitruvius and featured on the Horologion's reliefs.

But when I design a larger structure, such as a skyscraper, the numbers on the wind maps no longer hold. Wind is not linear: it doesn't change in a predictable way the higher you go into the atmosphere. Trying to extrapolate the data, or using mathematical trickery to adjust the numbers for 100-metre towers to fit 300-metre towers, will only produce unrealistic results. Instead, the structure has to be tested in a wind tunnel.

When I was working on the design of a 40-storey tower near the Regent's Canal in London, I visited one such facility. The miniaturised world of the wind-tunnel testers is a marvel in itself. In Milton Keynes, modelmakers had created a scaled replica of my

Horologion of Andronikos Kyrrhestes (Tower of Winds) built in the 2nd–1st centuries BC in Athens, Greece.

building that was 200 times smaller than the real thing would be. Not only that, they had also created tiny versions of all the other structures in the area, and the whole model sat on a turntable. The structures around my building were crucial to the data. If my tower was in the middle of a field, it would be hit directly by the force of the wind, unimpeded by any other object. In the middle of a metropolis, however, the densely textured cityscape with its mix of different structures affects the wind flow and turbulence, so the forces my building feels would differ.

I stood behind the model of my building and peered down the 'tunnel' – a long, square, smooth-walled passageway – towards the gigantic fan at the other end. It was set at the wind speed the building would feel from that particular direction. Once the cables connected to the apparatus were checked and the operatives ready, the fan was switched on. I braced myself as the blades whirred and a blast of chilly air shot through the miniature city in front of me, and hit me in the face. Inside the model of my building, thousands of sensors detected how much they were being pushed or pulled, and sent the numbers to a computer. The turntable was rotated by 15 degrees and the process repeated until the system had logged data from 24 directions. Over the next few weeks, engineers at the facility organised the data and prepared a report. I entered their numbers in my computer model to test my building. It was imperative that my structure remained stable against all the different effects that the wind could have on it, in every direction.

There are three ways in which wind can adversely affect a structure. First, if the structure above ground is light, wind can make it topple over, like the scattered traffic cones you see after a storm. Second, if the ground is weak, wind can cause the building to move and sink. Think of a sailboat on a windy day. The strength of the wind pushes the boat across the water – which of course is the desired effect if you're out sailing. But you wouldn't want your building or bridge to move sideways in the soil as the wind hits it. Now, soil is not as fluid as water, so you wouldn't see a building floating past you in a storm (if you do see this, take my professional advice and run the other way). But soil can still be squashed and moved around, so engineers need to provide an anchor – foundations – to keep their buildings in place.

The third effect is similar to a boat rocking at sea. Like trees, all buildings sway back and forth in the wind, depending on how strongly it is blowing – this is normal and safe. Unlike trees, however, buildings don't move so much that you can easily see the displacement. Towers are generally designed to bend through a maximum distance of their height divided by 500 – so a 500 m-tall tower won't move more than 1 m; but if this sway happens too quickly it could make you feel seasick.

One way to prevent a structure from toppling over is to make it heavy enough. In the past, most buildings were relatively modest in height and, because they were made from stone or brick, contained enough weight to resist the threat of the wind. But the higher you build, the stronger the wind is that you encounter. In the twentieth century, as we began to build taller and lighter structures, the force of the wind became a force to be reckoned with.

And so, in the modern skyscraper, weight alone is not always enough to keep it upright. Instead, the engineer must find a way to make the structure stiff enough to resist the wind. If you've ever watched a tree bending in a high wind and seen how it's able to withstand such a force, then you already understand the principle engineers use to keep modern buildings upright, even if it's blowing a gale outside. Just as a tree's stability depends on a solid, well-rooted but pliable trunk, so a building's stability often depends on a *core*, made from steel or concrete.

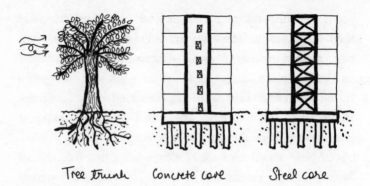

Tree trunk   Concrete core   Steel core

The core of a building, whether it's concrete or steel, is designed to provide the stable 'trunk' of the structure and so must be well-rooted in the ground.

The core – which, as the name suggests, tends to be in the centre of a tower – is an arrangement of walls in a square or rectangle that extends vertically throughout the height of a tower – like the spine in the human body. The floors of the building are joined to the core walls. The reason we don't generally notice cores is because they are well hidden, and usually themselves hide the essential services that are needed, like elevators, stairs, air ventilation ducts, electricity cables and water pipes.

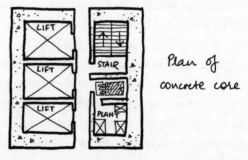

Plan of concrete core

Arranging the core of a building, usually hidden within the centre of the structure, which in turn provides a suitable place for essential services.

When wind hits the building, its force is channelled into and through the core. A building's core is a cantilever – a structure, like a diving board, that is clamped firmly at one end and free to move at the other. The core is designed to flex a little and allow the wind forces to flow down into the foundations, stabilising the core and the building – much as a tree's roots help it to withstand and disperse the wind's power.

The walls in a concrete core will be made of solid concrete (apart from holes in specific places for elevator or stair doors), making the core inherently stiff. Steel cores are different: simply replacing concrete walls with steel ones would be incredibly expensive and heavy; the sheer weight of the steel would make them impossible to build. So instead of solid walls, steel columns and beams are arranged in formations of triangles and rectangles to create a frame or vertical truss.

The force in each steel section or concrete wall depends on which direction the wind is blowing. My computer model has the wind force values for 24 different directions from the wind tunnel report. The forces create compression and tension in the beams, columns and struts that make up the frame in a steel core, or the walls in a concrete one. The computer then works out the compression and tension in every bit of the core, for every orientation. We then design each steel section or concrete wall using the largest compression and tension figures. We vary the size of the steel, or the thickness of the concrete, depending on the force in each. The core thus keeps the tower stable irrespective of wind direction. It's a complicated procedure to check the force in just one area for 24 different wind effects, let alone an entire core. Fortunately, computing power nowadays does the hard work, making it somewhat simpler for the engineer.

*

The building at 30 St Mary Axe in London, which is 41 storeys tall and shaped like a gherkin (hence its nickname), has a different way of remaining stable in the face of wind. The elegantly curved cylinder of shaded blue glass is surrounded by large criss-crossing pieces of steel in the shape of big diamonds.

Completed in 2012, 30 St Mary Axe, London – otherwise known as 'the Gherkin' – has a steel exoskeleton to protect it from external forces.

A core is like a spine or a skeleton, giving a building integrity from the inside, but 30 St Mary Axe is surrounded by an exoskeleton. This exoskeleton – or, to use the technical term, *external braced frame* or *diagrid* – is like the shell of a turtle. Instead of an internal structure that resists the forces trying to push it over, it's the shell or frame around the building that does the protecting. As it is buffeted by wind, the network of steel that forms the diagrid transmits the wind force to the foundations to keep the building stable.

Another spectacular example of the external braced frame is the Centre Pompidou in Paris. Architects Renzo Piano, Richard Rogers and Gianfranco Franchini envisioned what is, in effect, an inside-out building. All its arteries – the stuff that's usually hidden away, like fresh-water and waste pipes, electricity cables, ventilation ducts, and even the stairs, elevators and

The Pompidou Centre, Paris, has an external braced frame composed of a network of steel rods.

escalators – are on the outside of the building. It's these details that catch the eye and which people remember: the snaking pipes painted white, blue or green; the translucent tube of the escalator zigzagging upwards. But take a second look and you'll notice that the whole structure is clad in a network of large X-shaped rods, which are there to keep it stable against the wind. An exoskeleton, among the air ducts and waste pipes.

As a structural engineer, I like that I can see how the building works, and understand where the loads are going. Instead of hiding or disguising all the seemingly unglamorous but essential systems that make a building run smoothly, exposed systems like the Centre Pompidou's are delightfully honest, and treat us to an insight into the character of a structure.

*

Diagrids and cores, however, are not incorporated into buildings just to stop them toppling over – they also control sway. It might seem strange that our seemingly solid structures, made from steel and concrete, move – but they do. The swaying in itself is not a problem: what's important is how fast the building sways, and for how long. Through years of experiments we've been able to determine the levels of acceleration (a measure of how quickly the speed of an object is changing) at which humans can feel this movement. Take travelling in an aeroplane, for example: even though it flies extremely fast, in calm air you hardly feel you're moving at all. When you hit turbulence, however, the speed starts to change suddenly and quickly, and you feel it. Buildings are similar: they can move by quite a large amount, and you won't feel anything so long as the acceleration is small. But if the acceleration is large, then even if the building is only moving a small amount you could feel queasy.

It's not just the acceleration that affects us. How long the building continues to sway – how long it oscillates or moves side-to-side – can also make us feel unsteady. To use a diving-board analogy once more: when you bounce on the board and take a dive, the board oscillates before it stops moving. A thick board that is strongly clamped at its end only oscillates a short distance and stops after just a few oscillations. A thinner, weaker board that isn't as strongly clamped will oscillate a greater distance and for a longer time.

When I design a tall tower, I have to make sure that the acceleration of the sway is outside the range of human perception, and that the oscillation stops quickly.

The same computer model that helps me design a structure that can resist gravity and wind also helps me with this challenge. I enter the materials, shape and size of the beams, columns and core into the programme. The software then analyses the wind force, the materials' stiffness and the geometry of the structure, and tells me what the acceleration is. If it is below the threshold that people can feel, then nothing more needs to be done. If, however the acceleration is greater, then I need to make the structure stiffer. We can achieve this by increasing the thickness of the walls of a concrete core, or using bigger steel struts in a steel one. I then rerun the model, sometimes many times, until the target acceleration is reached.

The taller and more slender the tower, the more pronounced the sway. Sometimes it isn't possible to stiffen the structure enough to control the acceleration and how long it oscillates. So although the building is perfectly safe, it wouldn't feel safe. In that case, the sway of the tower is artificially controlled using a form of pendulum called a *tuned mass damper*, which moves in the opposite direction to the tower.

Every object, including buildings, has a natural frequency: the number of times it vibrates in one second when it is disturbed. An opera singer can shatter a wine glass because the glass has its own natural frequency. If the singer can hit a note with the same frequency as the glass, the energy of her voice causes the glass to vibrate dramatically until it rips itself apart. Similarly, wind (and earthquakes) can shake buildings at a particular frequency. If the natural frequency of the building is the same as that of the gusts of wind or the earthquake, the building will vibrate dramatically, and will be damaged. This phenomenon – an object vibrating dramatically at its natural frequency – is called *resonance*.

A pendulum – which is basically a weight suspended by cables or springs – oscillates back and forth. Depending on the length of the cable, or the stiffness of the springs, it swings a fixed number of times in a fixed period. When using a pendulum to cancel out a skyscraper's sway, the trick is to calculate the skyscraper's frequency (using a computer model), and

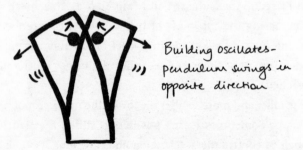

Building oscillates – pendulum swings in opposite direction

A pendulum cancels out the sway of a tall building by swinging in the opposite direction.

then to install a pendulum with a similar frequency at the top. When wind or an earthquake hits the skyscraper, it starts to move back and forth. This causes the pendulum to oscillate as well – but in the opposite direction to the tower.

You can stop the vibration of a tuning fork – and therefore its sound – just by touching one of its prongs. Your finger absorbs the energy of the vibration. The same process is at work in our

Standing at 509 metres tall, the Taipei 101 tower dominates the skyline of Taipei City, Taiwan.

swaying skyscraper. The building is like the tuning fork and the pendulum acts like your finger, absorbing the energy created by the movement of the skyscraper, which moves less and less. The movement of the structure is said to be 'damped' (hence the term 'tuned mass damper'), so the people inside can't feel it.

Taipei 101, the 509 m tower in Taipei City in Taiwan, was the tallest building in the world when it was completed in 2004. It is deservedly famous for its distinct architectural aesthetic: inspired by pagodas and stalks of bamboo, the building is composed of eight trapezoidal sections that give it a ridged, organic feel, as though it has pushed its way out of the ground like the stem of a plant – an illusion reinforced by the tinted windows, which give it a green hue.

But the tower is also famous for the huge ball of steel that hangs between the 92nd and 87th floors. At 660 tonnes, this steel pendulum is the heaviest in any skyscraper in the world. It is a huge

The pendulum in Taipei 101 is how the building survives earthquakes.

tourist attraction (its sheer scale, geometrical elegance and bright yellow colour make it look like something from a sci-fi film), but its real purpose is to protect the tower from the typhoons and earthquakes that can hit the city. When the building is shaken by a storm, or by an earthquake vibrating the ground beneath it, the pendulum swings into action, oscillating to absorb the movement of the tower. In August 2015, Typhoon Soudelor swept across Taiwan, gusting to at least 170 km an hour, but Taipei 101 escaped undamaged. Its saviour, the pendulum, recorded movement of up to 1 m – its largest ever movement.

*

Engineers use a pendulum to defend against wind and earthquakes because both are random forces that act in a horizontal direction. But earthquakes can have far more devastating effects, so we often need other precautions too. The terrifying, annihilative power of the earthquake gave rise to all manner of explanations for its origins. Ancient Indian mythology says that the Earth shakes when the four elephants that carry it on their backs move or stretch. According to Norse myths, the Earth trembles when Loki (the God of Mischief, imprisoned in a cave for his misdeeds) wrestles with his restraints. The Japanese blame the giant catfish, Namazu, which lives underneath the Earth in mud, guarded by a god who holds it down with a huge stone. Sometimes, however, the god becomes distracted and allows Namazu to thrash about. Nowadays we have a less colourful but more accurate explanation for the periodic vibration of the Earth. Earthquakes happen when different layers of the Earth's crust move relative to one another. A wave of energy explodes from a single point: the epicentre. The energy spreads outward from this point, shaking everything on the surface, including our structures. The waves of energy from the tremors that affect our structures are unpredictable and irregular – they strike without warning.

Engineers study the frequencies of earthquakes in historical records, then they use a computer model to compare these to the natural frequency of the building to be constructed. Just like we did for wind, we must ensure that the two frequencies aren't too similar, otherwise the building will resonate and could be damaged, or even collapse. If they are, the natural frequency of the building can be changed by adding more weight to it, or by making the core or frame of the structure stiffer.

Another way to mitigate the effects of an earthquake's energy waves is to use special rubber 'feet' or 'bearings'. If you sit in your living room with powerful speakers busting out some bass, you feel vibrations transmit from the speakers, into the floors, through the sofa and finally into your body. Put some rubber feet on the underside of the speakers and the effect lessens, because the feet absorb most of the vibrations. Similarly, we can install big rubber bearings at the bottom of the columns of a building, which then absorb an earthquake's vibrations.

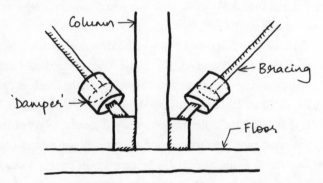

Dampers protecting the Torre Mayor skyscraper, Mexico City, Mexico.

Earthquake energy can also be absorbed in the connections between beams, columns and diagonal braces. The Torre Mayor skyscraper in Mexico City employs a very clever system to do this. In this 55-storey structure, 96 hydraulic dampers or shock absorbers – like pistons in a car – are arranged in X shapes all around the building and across its full height (creating a diagrid) to act as extra bracing against earthquakes. When an earthquake occurs, the whole building sways and the movement is absorbed into these dampers so the structure itself doesn't move too much. In fact, very soon after the Torre Mayor was completed, an earthquake recorded at a magnitude of 7.6 shook Mexico City, causing widespread damage. The Torre Mayor building survived unscathed; it's said that the occupants did not even realise there had been an earthquake.

And this, in a way, is the engineer's ideal – a building so well-designed, and so secure, that its occupants carry on comfortably with their business, completely unaware of the amount of complicated technology tackling all the forces the structure has to withstand each day.

FIRE

On the morning of 12 March 1993, I went to school in the Juhu district of Mumbai as usual, with my hair tied neatly back, wearing a crisp white blouse and grey pinafore. My teeth were hidden by braces, which were interwoven with my choice of green bands; definitely not cool (yes, even at nine I was already the class nerd). At 2.00pm Mum picked up my sister and me in our lime-green Fiat and took us home. While she was parking the car, we raced up four flights of stairs in our daily competition to see who could make it to our front door first. But something felt different. We stopped at the last step; we couldn't get to the door because our neighbour was standing there, nervously fiddling with her *dupatta*, looking distressed.

We soon discovered why. While Mum was collecting us from school there had been a bomb attack on the Bombay Stock Exchange – the building where my father and uncle worked.

Panicking, we ran into the flat and switched on the television. Every news channel was covering the mayhem. Bombs continued to explode around the city. Hundreds had been killed and injured. This was before the advent of mobile phones, so we had no way of knowing if my father and uncle were alive and safe.

The Bombay Stock Exchange is a 29-storey concrete tower in the heart of Mumbai's financial district. A car carrying a bomb had made its way into the basement garage and then detonated. Many lives were lost; many more people were hurt. I stood in front of the television horrified, watching images of weeping people covered in blood and dust running from billowing smoke. Police cars, fire trucks and ambulances raced to the tower, sirens blaring. We could see that the offices on the ground and first floor nearest to the explosion had been destroyed. It was clear that no one in that part of the building could have survived. Dazed people from the higher floors clambered down stairs and out of the tower. At home, we looked at each other and didn't utter a word, but I knew the same thought was running through all our minds. My dad and uncle worked on the eighth floor. We quietly hoped for the best.

As I learned later, my dad had been sitting at his desk, shouting down a poor phone line to one of his clients when a huge bang rattled the building. At first he thought an electricity generator or a large cooling unit had exploded. He jumped out of his seat, telling his staff to stay calm and remain in the office. Seconds later, however, he heard terrified people running down the stairs. Many screamed that there had been a bomb and that everybody should get out as quickly as possible. My father, uncle and their colleagues left their office, to scenes of horror.

Hundreds of people were filing down the stairs. There was barely any space to move. Head down, he focused on taking one step at a time, trying not to look at the dismembered bodies – the arms, the legs, the blood – that lay just beyond the staircase. Finally, he arrived at the ground floor. Emergency vehicles, trying to deal with the injured, blocked the street. My father and uncle fled the area and got on a bus to my grandmother's house. About two hours after we'd come home from school – the longest two hours of my life – Pop called us to tell us they were both safe.

Years later, while studying for my masters in structural engineering, I attended a class in which we discussed how to protect towers against explosions. Suddenly, the events of that terrible day in March came rushing back. For the first time a thought occurred to me: given that it was rocked by a serious explosion right at the base of the structure, and fires broke out afterwards, why didn't the whole Bombay Stock Exchange tower collapse?

I know now that there are two main reasons for this. The first is that engineers design certain buildings to resist explosions, so even if it is hit and damaged, it doesn't collapse like a house of cards. There is a minimum standard of safety governing the design of *all* structures, but the more vulnerable ones – tall, iconic buildings, for example, or those with particularly large numbers of people inside – are designed specifically for a range of possible explosion scenarios. The second reason is that all structures should be designed to stop fires rapidly engulfing them, providing enough time for occupants to escape, and for the fire to be tackled or burn out – contained in a small area – before it causes significant structural failure.

But we didn't start out building this way; we have learned from disasters of the past.

*

After waking early on the morning of 16 May 1968, Ivy Hodge went to the kitchen to make a cup of tea. She turned on the gas hob, struck a match – and the next thing she knew she was flat on her back, looking at the sky. A wall of her kitchen and a wall of the living room had disappeared.

In Ivy's flat on the 18th floor of a 22-storey tower block in Canning Town, London, there had been an explosion. Occurring in peacetime in a quiet residential neighbourhood, it was an event without precedent in the city, and it profoundly influenced how we would build future structures.

The tower had been constructed quickly as part of the regeneration desperately needed in the aftermath of the Second World War. The neighbourhood had lost about a quarter of its homes to bombing, and the destruction, coupled with the large post-war population increase, meant there was a severe housing shortage. To build rapidly and efficiently, new forms of construction were being experimented with. This particular structure was the second of nine identical towers being built to create an estate called Ronan Point.

The tower had been thrown together hastily by 'pre-fabrication'. Instead of pouring wet concrete on a construction site and waiting for it to solidify to form walls and floors (like most other concrete construction required), room-sized panels of concrete were made in a factory. The panels were then driven to site and lifted in to place with a crane. It was like building a house of cards: put up the walls of the ground floor, carefully place the horizontal panels on top of them to create the first floor, and so on, up and up. The panels were joined together with a small amount of wet concrete on site. The weight of the building was being channelled through these large load-bearing panels; there was no skeleton or frame. This novel prefabricated system produced lower costs, quicker construction times and required less labour, all important economic factors to consider in recovering post-war Britain.

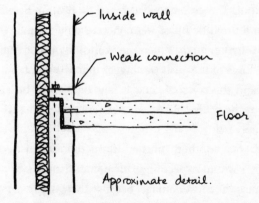

Poor detailing, such as that used at Ronan Point, where only a small amount of wet concrete was used to join together prefabricated panels during construction.

In Ivy Hodge's flat, gas had been leaking steadily from her recently installed but defective boiler system. The match flame had lit the escaped gas and BOOM!, the wall panels making up the corner of her flat blew out. With nothing now supporting them, the wall panels of the flat above fell, hitting the level below. One by one, each floor on that corner of the towerblock collapsed, taking a great chunk out of the structure, from top to bottom. Four people, asleep in their flats, died.

Oddly, the explosion did not perforate Ivy's eardrums, which suggests that its force wasn't that large – since it doesn't require much pressure to damage them. In fact, subsequent investigations showed that even an explosion with just a third of the force of the actual event would have dislodged the wall panels. Since the panels were just sitting one on top of the other, without being tied together properly, there was little to stop them blowing out. The designers had relied on friction between the panels and the little bit of wet concrete 'glue' to hold them in place. It wasn't enough. When the explosion pushed out on the wall, the force of the push was bigger than the resistance of the friction and the concrete, and it flew out. Then, because the load from the walls above had nowhere to channel itself, the walls simply fell.

There was another unusual thing about this collapse. Normally I would expect an explosion at the base of a building to cause the most damage, because there are many storeys above it which can come crashing down. In this case, however, if the same explosion had happened at the base of the building, the collapse might not have happened at all.

Friction depends on weight. The heavier the load acting at the junction between two surfaces, the greater the friction. Close to the top of the tower (where Ivy was), there were only

The disproportionate collapse of floors following an explosion at Ronan Point, London, in 1968.

four storeys of weight at the junction between wall and floor, so the friction was low. The pressure of the explosion overcame the friction and sent concrete panels flying. But at the base of the tower, the weight of more than twenty storeys of panels created greater friction between wall panels (it's the reason why pulling a magazine out of the base of a stack is much harder than extracting one from higher up). So counter-intuitively, the explosion near the top was the event with disastrous results. This is not a very common occurrence now – especially because, as we'll see, buildings aren't built like this anymore.

The debacle at Ronan Point had two important lessons for future construction. Firstly, it was vital to tie structures together, so that if a wall or floor panel were pushed with a force bigger than expected, the ties would stop the panels from sliding out. (At Ronan Point, steel rods, for example, tying together the prefabricated wall panels between floors could have helped the building withstand the blast; variations of this tie-system are used in modern prefabricated buildings.) Even for structures built in a more traditional way, with all the concrete poured, or steel being fixed on site, it is essential to make sure that the beams and columns have robust connections. In the case of steel frames, the bolts used to join pieces of steel together should be strong enough not only to resist normal loads exerted by wind and gravity, but also to keep the structure bound together.

Secondly, engineers had to prevent disproportionate effect. At Ronan Point, a single explosion on the 18th floor caused the corner of the tower to collapse at all levels. This domino-effect was disproportionate to the cause, and a new term, *disproportionate collapse,* was born. If an event like an explosion happens, then of course damage will occur, but the effect of an explosion on one storey shouldn't propagate throughout the structure. The problem at the Canning Town tower block was that the loads didn't have anywhere to go. So the key is to ensure that the forces have somewhere to go, even if part of a structure disappears. It's like sitting on a stool: theoretically, only a quarter of your weight is transmitted through each of the four legs. But if, like many people, you're inclined to tilt the stool so all your weight is going down only two legs, you've just doubled the load the leg is designed for – the legs fail, you hit the ground, and you bruise your backside. But if structural engineers anticipate this sort of behaviour and design every leg for double the load, then you're safe.

Thus the idea of consciously creating new paths for loads to travel through was born. In my computer model, I will delete a column, record the larger forces in neighbouring columns, and design for this higher load. Then I know that even if that column is gone, its neighbours will do its job. Then I put that column back in and remove another one, trying different combinations to check my structure is stable in the face of explosions. Never challenge a structural engineer to a game of Jenga: we know which blocks to remove – how to take chunks out of a structure so that it doesn't crash.

*

Throughout history, engineers and civic authorities have been engaged in battle – against the fires that threaten to raze our towns and cities to the ground. Roman houses were often made with timber frames, floors and roofs, which caught alight easily, and fires were common. The Great Fire of Rome in AD 64 laid waste to two-thirds of the city. Originally, timber was not protected with anything to resist fire like it is now, and walls were made from wattle and daub. Wattle, a lattice woven from narrow wooden strips which looked a bit like a straw basket, was coated – daubed – with a mixture of wet soil, clay, sand and straw. Such a construction was highly flammable, enabling fire to spread quickly. The narrow streets aggravated the situation because flames could easily jump the small distance between one building and another.

In the first century BC, Marcus Licinius Crassus was born into the upper echelons of Roman society. He grew up to become a respected general (he helped quash the slave revolt of Spartacus) and a notorious businessman. Crassus was a man who spotted opportunities: observing the devastation caused by Rome's fires, he created the world's first fire brigade, made up of over 500 slaves who were trained to fight fires. He ran it as a private business, rushing his team to burning buildings, where they intimidated and drove away rival firefighters, then stood about until Crassus had negotiated a price to put out the fire with the building's distraught owners. If no deal could be reached, the firefighters simply allowed the structures to burn to the ground. Crassus would then offer the owners a derisory sum to purchase the smoking site. This meant that he quickly managed to buy up much of Rome, and amassed a fortune as a result. Fortunately, modern-day fire brigades work on a more honest basis.

After the Great Fire of Rome, Nero ordered several changes to the city. Streets were made wider, apartment buildings limited to six storeys, and bakers' or metal workers' shops separated from residential units, using double walls with air gaps. He proclaimed that balconies should be made fire-proof to make escape easier, and invested in improving the water supply, so it could be used to extinguish fires. The Romans learned from tragedy, and we too have benefited from that hard-won wisdom. Thousands of years later, these simple principles – separating rooms, flats and buildings with fire-resistant materials and installing air gaps – are still used to prevent fires ravaging modern structures.

*

On 11 September 2001, the world watched in horror as two planes collided with the World Trade Center towers in New

York. I was in Los Angeles on holiday before starting at university, and was scheduled to fly to New York the next day. Paralysed, I sat watching the news, shocked as the towers collapsed an hour after being hit. A few days later, I went directly back to London, already feeling part of a changed world.

Looking at the events from an engineer's point of view, the events of that appalling day had a ripple effect on the design and construction of skyscrapers. Reading about the structural failures that led to the collapse of the towers, I was surprised to learn that it wasn't just the impact of the planes that caused the devastation, it was also the fires that followed.

New York is filled with spectacular skyscrapers, yet the World Trade Center's twin towers (opened in 1973) were among the city's most iconic symbols. Visually, each of the towers was very simple – a perfect square from a bird's-eye view, 110 storeys high. Each had a large central core made of steel columns. But this spine wasn't responsible for keeping the towers stable: they used the 'turtle-shell'-style exoskeleton instead.

Vertical columns, spaced just over a metre apart all around the perimeter of the square, were joined up at each storey with beams. The beams and columns together formed a robust frame, similar to the construction of the Gherkin we saw earlier, but with giant rectangles instead of triangles. The connections between the beams and columns were very stiff. This external frame kept the building strong against the force of the wind.

When the planes crashed into the towers, giant holes opened up in the exoskeleton. A number of columns and beams were destroyed. Engineers had in fact planned for the possibility of some form of impact by aeroplane. They had studied what might happen if a Boeing 707 (the largest commercial aircraft in operation at the time of construction) hit the building, and they had designed accordingly. The beams and columns had been constructed with extra-strong connections tying them together, so even though some of the structure was gone, the loads found somewhere else to go: they flowed *around* the hole (using the principle of preventing disproportionate collapse, which engineers had learned from Ronan Point).

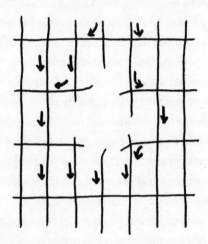

Loads within a building find new routes as the forces are channelled through alternative load paths.

The planes that hit the twin towers were not the Boeing 707s that engineers had planned for nearly 30 years earlier; they were larger 767s, carrying more aviation fuel. On impact, the fuel caught fire, and the conflagration of the fuel, aircraft parts, desks and other flammable material inside the building made the steel columns very hot. When steel gets hot, it behaves badly: the tiny crystals which make up the material become excited, vibrate and begin to move around, and the normally strong bonds between them are loosened. Loose bonds mean soft metal. So hot steel is weaker than cold steel, and cannot bear the same load. On 9/11, the columns just next to the holes were supporting a larger load than usual, because they were channelling not just their own forces but also those their neighbours had once carried. The steel columns and floor beams had been sprayed with a special paint mixed with mineral fibres, designed to insulate the steel from the heat of a fire and prevent it from getting too hot. But the crash of the plane and the projectile debris had chipped away areas of the protective paint, leaving big patches of exposed steel. The temperature of the columns around the perimeter of the tower rose ever higher.

The steel columns which made up the core also became unnaturally hot. Two layers of gypsum board (a panel made of gypsum plaster pressed between two thick sheets of paper) separated the core from the rest of the building. The idea was that a fire in the office space couldn't infiltrate the core past these boards, so people could run into this safe zone and to the stairs to escape. But this board was damaged, leaving the core columns susceptible and the intended safe passage exposed.

The columns became weaker and weaker, and as temperatures reached about 1,000° Celsius, they gave up. They couldn't carry the forces any more and they bowed.

In the end, the columns failed completely and the structure above it was then left vulnerable to the effects of gravity. The floor above the failed columns came crashing down. But the level on which it landed wasn't strong enough to resist the falling load and it too failed. One after the other – in a domino-effect reminiscent of the Canning Town disaster but on an even more shockingly huge scale – all the floors failed and the towers came down. The fire protection – paint and boards – was no match for the size and intensity of the fire.

The way we design skyscrapers has changed since that day. Now, we make sure that escape routes are protected more robustly. The easiest way to do this is to build the core in concrete instead of steel, so that instead of weak gypsum boards standing between the fire and safety, you have a solid wall of concrete.

Concrete is not a good conductor: it doesn't transmit heat well, which means it takes longer to heat up. To strengthen concrete, however, we insert steel reinforcement bars into it; these are excellent conductors of heat, which creates a problem for the engineer. In a fire the steel bars heat up, and the heat energy spreads quickly through their length, while the surrounding concrete heats up slowly. The hot steel expands more quickly than the colder concrete, causing the outer layers of concrete to crack and burst off. This is similar to how thick glass tumblers crack if you pour hot water into them: the inner layers of the glass get very hot and expand, but the outer layers remain cold because glass, like concrete, is not a good conductor of heat. As the inner layers expand against the colder outer layers, the outermost cracks.

Through testing and experimentation, we know how long it takes for concrete to conduct heat to steel bars, and then for the steel bars to heat up and make the concrete burst. So we bury the steel deep enough in the concrete to ensure that the fires can be put out before the outer layer of concrete is damaged. This buys enough time for people to leave the building through the concrete core, or for firefighters to get the flames under control, without the structure collapsing. The taller or larger the building, the longer it takes to escape, so the deeper the steel is embedded in the concrete. Just a few centimetres make a tremendous difference.

So concrete cores perform a dual function: keeping the building stable against wind loads, and forming a protected escape route for the occupants. Today, even if we use an exoskeleton to resist wind (which means we don't *need* an internal core), we still often install concrete walls to safeguard escape routes. And the protection for steel columns and beams against fire has also been improved dramatically: fire-resistant boards and intumescent paint (which expands when heated and insulates the metal) are much more robust now than ever before. They stop steel getting too hot too quickly, so it remains strong.

Learning from disasters is fundamental to engineering: part of the engineer's job is a constant process of improvement, endeavouring to build structures that are better, stronger and safer than they were before. Thanks to such lessons we now anticipate the removal of columns, and check in advance that a building will not collapse. The Bombay Stock Exchange tower had been built in such a way that even though the structure in the immediate vicinity of the car bomb was severely impacted, the loads it was carrying found somewhere else to go. The damaged part of the building remained stable enough because it was tied into the rest of the structure, so – unlike Ronan Point – the floors above didn't come crashing down. The steel bars buried in the concrete walls and columns held their strength in the face of the fires that blazed after the explosion.

It was the lessons engineers learned from history, and the new way of designing for the unanticipated, which saved my dad's life that day.

CLAY

I love baking, which is perhaps not surprising, given that it has a lot in common with engineering. I like the way you have to follow an ordered series of processes to construct a cake. I like that you work in a very patient and precise fashion, otherwise you won't get the right shape and texture. I like the hopeful wait, that quiet period when my work is done and it slowly takes shape in the oven. Usually, I find all this incredibly satisfying. But there are moments of perplexed frustration – like the time I opened the oven door ready to slide out a delicious pineapple upside-down cake and was confronted instead with chunks of uncooked fruit swimming listlessly in a greasy sea of butter. Forget soggy bottom, this was a soggy disaster. Cursing the oven and recipe (after all, it could hardly have been my fault), I slung it straight in the bin: useless – except as a valuable reminder that in baking, as in engineering, the right choice of materials, combined in the correct way, is crucial to the outcome.

When designing a building or bridge, materials are one of my foremost concerns. In fact, different materials can entirely change the way the frame of a structure is arranged, how intrusive it feels, and how physically heavy and expensive it is. They must serve the purpose of the building or bridge correctly: I need to weave in the skeleton of the structure without it becoming obtrusive to the people using it. The materials must also resist the stresses and strains of loads that assail a building, and perform well in the face of movement and temperature fluctuations. Ultimately, my choice of material has to ensure that the structure survives as long as possible in its environment. Luckily, my engineering creations are more successful than my baking endeavours.

The science of materials has long obsessed humans, and since ancient times we have theorised about what makes up 'stuff'. The Greek philosopher Thales (c. 600 BC) contended that Water was the primordial substance of all things. Heraclitus of Ephesus (c. 535 BC) said it was Fire. Democritus (c. 460 BC) and his follower Epicurus suggested it was the 'indivisibles': the precursors to what we now call atoms. In Hinduism, the four elements – earth, fire, water and air – described matter, and a fifth – *akasha* – encompassed that beyond the material world. Roman engineer Vitruvius writes in *De Architectura* agreeing that matter is made up of the same four elements, adding that the behaviour and character of a material depends on the proportions of these elements within it.

This idea – that there were a limited number of fundamental ingredients which in different proportions could explain every colour, texture, strength and other property of any material – was revolutionary. The Romans surmised that materials which were soft must have a larger proportion of air, and that tougher materials had more earth. Water in large proportions made a material resistant to it, and brittle materials were ruled by fire.

Ever curious and inventive, the Romans manipulated these materials to better their properties, which is how they made their renowned concrete. They may not have had the periodic table (it would be a while before Dmitri Mendeleev published the original version of the table in 1869), but they knew that the properties of a material depended on the proportions of its elements, and they could be changed by exposing it to other elements.

For a long time, however, humans simply built from the materials that Nature provided, without changing their fundamental properties. Our ancient ancestors' dwellings were made from whatever they could find in their immediate surroundings: materials that were readily available and could be easily assembled into different shapes. With a few simple tools, trees could be felled and logs joined to create walls, and animal skins could be tied together and suspended to form tents.

If there were no trees, humans created homes from mud. As we developed our tools and became more innovative and daring, we took this one step further – we tried to make the mud better by shaping it into rectangular cuboids of various sizes using wooden moulds. We discovered that by allowing the mud to dry in the sun (according to Roman philosophy, letting the water escape and the earth take over, using fire), the result was a much tougher unit. Humans had created the brick.

Bricks were already in use around 9000 BC in an expanse of desert in the Middle East. In the deep valley of the River Jordan, hundreds of metres below sea-level, Neolithic man created the city of Jericho. The residents of this ancient city baked hand-moulded flat pieces of clay in the sun and built homes with them in the shape of beehives. As early as 2900 BC the Indus Valley Civilisation was building structures using bricks baked in kilns. It was a process that required skill and precision: if it wasn't heated for long enough, the shaped mud wouldn't dry out properly. Heated too much and too quickly, it would crack. But if baked at the right temperature for just the right length of time, the mud became strong and weather-resistant.

Archaeological remains from the Indus Valley Civilisation have been found in the ruins of Mohenjo-daro and Harappa, in modern-day Pakistan. Every brick they used, no matter what its size, was in the perfect ratio of 4 : 2 : 1 (length : width : height) – a ratio that engineers still (more or less) use, because it allows the brick to dry uniformly, it's a handy size to work with, and it has a good proportion of surface area that can be bound to other bricks with whatever form of glue or mortar is used. At about the same time as the Indus Valley Civilisation, the Chinese were also manufacturing bricks on a large scale. But for the humble brick to become one of Western civilisation's most used materials, we had to wait for the rise of one of its greatest empires.

*

The energy and inventiveness of Roman engineering is, for me, a source of wonder and inspiration. So it was with not a little excitement that I took a train south from Naples, along the coast, to one of the most famous archaeological sites in the world. Wearing matching sandals, my husband and I alighted at our destination and put on matching safari hats to keep the scorching summer sun at bay. In great anticipation, we strode towards the ancient ruins of Pompeii.

Along the cobbled streets were shopfronts with counters studded with holes in which conical pots or *amphorae* were once stored. On the ground was a dramatic floor mosaic of writhing fish and sea creatures. Another showed a ferocious canine and was inscribed with the legend 'Cave canem' – 'Beware of the dog'. Alongside these were well-laid-out homes, like Menander's (a Greek writer), with its spacious atrium, baths and garden surrounded by a beautifully proportioned colonnaded walkway or *peristyle*. All these gave a powerful impression of what a glorious, bustling town it must have been in its heyday.

Among the things that most caught my eye, though, were the blood-red bricks. They were everywhere. They peeked surreptitiously from columns on which the decorations that originally hid them from view had crumbled away. They looked proudly on from the walls, where they were arranged in thin layers of three, alternating with sharply contrasting layers of white stone. But my favourite brick-built features were without doubt the arches.

Arches are important building components. They are curved – they are a part of a circle or an ellipse, or even a parabola. They are strong shapes. Take, for example, an egg: if you squeeze an egg in your hand with a uniform grip, you'll find it nearly impossible to break because the curved shell channels the uniform force of your hand around itself in compression, and the shell is strong in resisting it. To crack the shell, you normally have to use a sharp edge, such as the blade of a knife, on one side, creating a non-uniform load. When you load an arch, the force is channelled around its curved shape, putting all portions of the arch in compression. In ancient times, stone or brick were commonly used building materials – these are great under these squashing loads but not tension loads. The Romans understood both the properties of such materials and the virtues of the arch, and they realised they could bring the two things together in perfect union. Until then, flat beams were used to span distances, whether in bridges or buildings. As we saw earlier, when loaded, beams experience compression in the top and tension in the bottom – and since stone and brick aren't very strong in tension, the beams the ancients used tended to be large and often unwieldy. This limited the length of the beams' spans. But by using the high compression resistance of stone in an arch, the Romans could create stronger and larger structures.

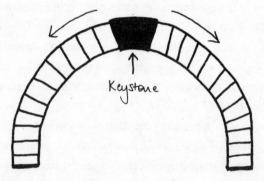

Forces channel around the curve of the arch; it is all in compression all of the time.

The brick arches surrounding me had survived millennia, and made me think of the beautiful ancient Arabic saying 'Arches never sleep.' They never sleep because their components are continuously in compression, resisting the weight they bear with endless patience. Even when Mount Vesuvius spewed lava over Pompeii, smothering its people and buildings, the arches remained the watchers of the city. They may have been buried, but they never stopped doing their job.

The ruins of Pompeii show us that the Romans used brick in almost every form of construction in the lands they conquered. In Italy and elsewhere, legions operated mobile kilns, spreading this practice as far as what are now the British Isles and Syria. You won't be surprised to learn that Vitruvius had an opinion on the material needed to make a perfect brick, the description for which he outlined in *De Architectura*. Creating a brick is much like creating a cake, so here's my take on a recipe for The Ancient Brick, courtesy of a range of ancient engineers – one that even I would be able to follow.

RECIPE FOR THE ANCIENT BRICK

Ingredients

Clay

'They should not be made of sandy or pebbly clay, or of fine gravel, because when made of these kinds they are in the first place heavy; and secondly, when washed by the rain as they stand in walls, they go to pieces and break up, and the straw in them does not hold together on account of the roughness of the material.

They should rather be made of a white and chalky or red clay, or even of a coarse-grained gravelly clay. These materials are smooth and therefore durable; they are not heavy to work with, and are readily laid.'

The waters of fruit

Warmth, in the form of the sun or a kiln

Method

1. Throw a lump of clay into knee-deep water and then stir and knead forty times with your feet.

2. Wet the clay with the waters of pine, mango and tree bark, and the water of three fruits, and continue kneading it for a month.

3. Form the clay, mixed with a little water, into large, flat rectangles using a wooden mould. (The Greek Lydian brick – typically used by the Romans, as per Vitruvius – is a foot and a half long and one foot wide.) Once formed, remove the bricks from the moulds.

4. Heat the clay gently and gradually. If made in the summer the bricks will be defective because the heat of the sun will cause their outer layers to harden quickly, while leaving the insides soft and vulnerable. The outer, drier layers will shrink more than the moist inner layers, causing the bricks to crack. On the other hand, if you make the bricks during the spring or the autumn they will dry out uniformly, due to the milder temperature.

5. After an interval of between two and four months, throw the bricks into water, take them out and allow them to dry completely.

Patience is key, as it takes up to two years for bricks to dry completely. Younger bricks will not have dried out completely, so may shrink over time. A wall made from such bricks and then plastered over will be seen to crack. Vitruvius alerts us to this: 'This is so true that at Utica in constructing walls they use brick only if it is dry and made five years previously, and approved as such by the authority of a magistrate.'

Roman bricks were, in general, larger and flatter than those we use today. They looked more like tiles: the Romans favoured that shape because they realised that, with the tools and methods they used, flatter bricks would dry out more evenly – an essential feature of the ideal brick recipe. From the temples in the Forum in Rome to the Colosseum and the monumental walls and arches of the Red Basilica in their ancient city of Pergamon, bricks formed the basis of their most impressive structures.

The ruins of the Red Basilica made from Roman bricks in modern-day Bergama, Turkey.

When the Roman empire fell in AD 476, the art of brickmaking was lost to the West for several hundred years, only to be revived in the Early Middle Ages (between the sixth and tenth centuries), when they were used to build castles. During the Renaissance and Baroque periods (from the fourteenth to the early eighteenth centuries), exposing bricks in buildings went out of fashion, and instead they were hidden behind intricate plaster and paintings. Personally, I like seeing bricks on display, much as I like seeing the air ducts and escalators on the outside of the Centre Pompidou. I prefer my structures direct and honest: like my cakes, I enjoy being able to view the materials from which they are created (this has nothing to do with my complete lack of icing skills).

During the Victorian period in Great Britain (1837–1901), and between the World Wars, the use of brick peaked to its highest in recent history. One of my favourite buildings in London, George Gilbert Scott's grand Gothic fantasy the St Pancras Renaissance Hotel, is a spectacular example of an exposed brick structure. Up to ten billion bricks were made annually in Britain. It seemed that all structures, from factories

The brickwork of a Roman arch at Pompei, southern Italy.

to houses, from sewers to bridges, were made from bricks, left exposed for all to see.

*

Such a timescale, stretching back millennia, is already hard to get your head around. But that's nothing compared to the dates involved in the creation of the raw material that makes up a brick. During the filming of *Britain Beneath Your Feet*, a two-part documentary about the ground and what's under it, I visited a clay mine in north-east London. There I was confronted by a vast clay cliff, sculpted by diggers from the ground on which London sits. The mine owner pointed to the top of it, which was the colour of rust. 'That clay is new, it's only twenty million years old.' My flabbergasted expression prompted him to continue. He explained that the 'newer' layers of clay had a much higher iron content, giving them a reddish tint. The stuff at the foot of the cliff was purer, so it had a blue-grey hue – a sure sign that it was older.

By 'older', he meant more than 50 million years old. Long ago, igneous (volcanic) rocks were weathered and transported by water, wind and ice. While the rocks and stones were being carried along, they picked up particles of other minerals such as quartz, mica, lime or iron oxide. This mixture of rock and minerals was deposited far from its original home in layers of sediment at the bottom of rivers, valleys and seas. In these environments, plants and animals thrived then died, adding a layer of organic matter that would then be covered by more rock, and so on. Gradually, over millions of years, under the right conditions of temperature and high pressure, these layers turned into sedimentary rock. And that's what the miners were busy digging out of the cliff face. The owner told me that, because of its incredible age, the clay is full of the fossils of tropical plants such as mangrove palms (which once flourished in British climes), and the ancestors of birds, turtles and crocodiles that no longer exist on Earth.

The mined clay is used for many things: crafting pots, art projects in schools and, of course, making bricks. For this, it is transported from the mine to factories where it is transformed into neat, solid cuboids. The principle of heating clay to create a brick hasn't changed from ancient times, but the method has. First, we treat the clay by adding extra sand or water to make it the right consistency: stiff but malleable. Then it is put into a machine that extrudes it through a mould or die (a bit like the hand-press in a Play-Doh Fun Factory, but on a slightly larger scale). The clay emerges in a long, rectangular column, which is chopped into brick-length pieces and conveyed to a dryer to gently remove as much moisture as possible – otherwise you end up with the cracked bricks that Vitruvius cautioned us about. The dryer is set at the relatively low temperature of 80–120° Celsius and is humid enough to stop the bricks from drying too quickly on the outside while the insides are still damp. And as they dry, they shrink.

If the process is stopped here, bricks similar to the ancient kiln-dried examples would be created. The next step is where the real difference between ancient and modern lies. The bricks are fired at temperatures of between 800° and 1,200° Celsius, fusing the particles of clay together so that they undergo a fundamental change. Clay turns into ceramic: more similar to glass than dried mud. This *fired* brick is far more durable than a *dried* brick, and that's what we use to build structures today. A fired brick is pretty strong: if we took the four elephants that support the Earth (and cause earthquakes when they stretch) from Indian mythology, added one more for luck, persuaded them to stand on top of each other and then tip-toe onto a single brick, the brick would remain intact.

To turn a single brick into a usable structure, we need a special glue or mortar that can bind the units together to form a whole. The ancient Egyptians used the mineral gypsum to make a plaster (also known as plaster of Paris, since it was commonly found and mined in the Montmartre district of the city). Unfortunately, however, gypsum isn't stable in the presence of water, so gypsum-sealed structures will eventually suffer damage and degradation. Fortunately, the Egyptians also used a different mixture that had lime mortars. This hardened and strengthened as it dried (and absorbed carbon dioxide from the atmosphere) and is more resilient than the gypsum recipe. When made correctly, mortars give strength to the structures they form and can last a very long time: parts of the Tower of London were built largely with lime mortar, and are still standing strong more than 900 years later.

Other materials are often mixed into the mortar to give it different properties. In China, the mortar used to build The Great Wall had a small amount of sticky rice added to it. Rice is mainly composed of starch – this made the mortar bond well with the stone, but also allowed some flexibility, so it wouldn't crack easily if the wall moved slightly as it heated and cooled with the seasons. The Romans added the blood of animals to their mortars, believing it helped the mortar stay strong when it was hit by frost. The dome of the Taj Mahal is held together with *chuna*, a mixture of burnt lime, ground shells, marble dust, gum, sugar, fruit juice and egg white.

Bricks are used in most UK houses today because they are cheap. But they have their disadvantages. You need specialist labour to lay the units one at a time, and it's a relatively slow process. And because of the standard size of the unit, you have less flexibility in the shapes of the structures you can create. Brick structures are also very weak in tension: the mortar glue between bricks, and the bricks themselves, can crack if pulled apart. Bricks can only be used in structures in which they are being compressed most of the time. They aren't strong enough to carry the weight of taller structures (steel and concrete can take far more compression than brick, as we'll see) so are impractical for, say, high-rise buildings or the larger bridges. However, their popularity remains where cost is the driver. Approximately 1.4 trillion bricks are made each year around the world; China alone manufactures about 800 billion, and India about 140 billion. LEGO, by comparison, makes a mere 45 billion or so bricks per year.

This ancient building block, born from the earth and baptised by fire, is so versatile that it was used in the construction of pyramids, the Great Wall of China, the Colosseum, the medieval Castle of the Teutonic Order in Malbork, the famous dome of the Catedrale di Santa Maria del Fiore in Florence, and even my own house. I love that in our modern, fast-paced world, with all the technology we've developed, we continue to rely heavily on a building tool that has been in use for over 10,000 years, created from a material that was 50 million years in the making.

METAL

In Delhi in India, there is a pillar of iron that doesn't rust. This column stands discreetly within the Qutb complex, a historic compound filled with extraordinary examples of Islamic architecture. The cavernous tomb of Iltutmish, in which every inch of the arched walls is decorated with loops and whorls, and the imposing Qutb Minar, a gracefully ridged, tapering tower – and at 72.5m the tallest brick minaret in the world – are simply breath-taking. At first glance, the dark grey column – about as thick as a tree trunk and barely seven metres tall – seems insignificant and out of place: a stray cat in a zoo of exotic animals. But it made a big impression on me.

The pillar predates the architecture around it. It was made in around AD 400 by one of the kings of the Gupta dynasty, as an offering to Lord Vishnu, the Hindu god worshipped as the Preserver of the Universe. Originally it was topped with a statue of *Garuda* (Vishnu's part-human, part-eagle steed, believed to be large enough to block out the sun). People consider it lucky if you can stand with your back to the pillar and wrap your arms around it so your fingers touch, but a fence now protects the monument from tourist limbs. I wasn't interested in luck, though, I was fascinated by another peculiar property of the pillar: in defiance of its natural propensities, this iron hasn't rusted in over 1,500 years.

The iron pillar that never rusts at the Qutb complex, Delhi, India.

The Iron Age followed the Bronze Age, which came to an end as copper and tin, the raw materials for making the metal, became difficult to obtain. The Iron Age is believed to have started around 1200 BC in India, and in Anatolia (modern-day

Turkey). Archaeologists studying the ruins of Kodumanal, a small village in the middle of Tamil Nadu state in southern India, found a trench dating back to around 300 BC on the southern edge of the village. In this was a furnace that still contained some iron slag (a by-product left over from the smelting of metals). Indian iron – mentioned in the writings of Aristotle and in Pliny the Elder's *Historia Naturalis* – was famous for its excellent quality. It was exported as far as Egypt for use by the Romans, but its secret recipe was carefully guarded.

To build the Iron Pillar, the ancient Indians made discs of iron, which they then forged (heated up and hammered together), before striking and filing the outer surface to make it smooth. The iron used to forge the column was extraordinarily pure, except for the higher than usual amounts of phosphorus it contained; a result of the extraction process used by the ironmongers. It is the presence of phosphorus that prevents the pillar from rusting. Rust forms on iron when it is exposed to oxygen and moisture; at first, the metal would have corroded but, in the dry local climate of Delhi, the phosphorus was drawn to the interface between the rust and the metal surface, creating a very thin film. This film prevents air and moisture from reacting with the iron. And so the pillar hasn't rusted any further. Modern steel is not made with those relatively high levels of phosphorus because the steel would become susceptible to cracking when it is 'hot-worked', which is a typical part of the manufacturing process where the metal is deformed at high temperatures. Take a look at structures made from iron or steel that are exposed to the atmosphere and you'll see they are painted to prevent the formation of rust, which would weaken them. But the steel beams and columns in our air-controlled buildings are left unpainted – unless painting is necessary for fire protection – because the lack of humidity means they won't rust much.

While the ancients recognised the wonders of iron, it was mostly used to make household vessels, jewellery and weapons, because the iron they extracted was too soft to build with, and they didn't know how to strengthen it enough to create an entire building or bridge. There are nonetheless rare examples of structures that use it: in *A Record of Buddhistic Kingdoms*, the Chinese monk Fa Hsien wrote about suspension bridges held up by iron-link chains in India around the time the pillar in Delhi was made. And the monumental marble gateway to the Acropolis in Athens, the Propylaea (built in around 432 BC), has iron bars to strengthen the ceiling beams. That's how the ancient engineers used metal: in little snippets to help strengthen their stone and brick structures. Before iron (or its cousin steel) could be used in large-scale structures, scientists and engineers had to learn more about its character.

*

Bricks and mortar crack easily when pulled apart, but metals don't. They are fundamentally different because of their molecular structure. Like diamonds, metals are made from crystals – but not large shiny ones like we see shimmering on the dresses of glamorous Bollywood actresses. Metal crystals are tiny – so small, in fact, that you can't see them with the naked eye – and they are opaque.

These crystals are attracted to each other, and this attraction bonds them together in a matrix or grid. However, when you heat up a metal, the crystals vibrate faster and faster until the bonds weaken. The metal then becomes malleable, and may even melt into a liquid if the temperature is high enough. Because of the flexibility of the bonds, metals are *ductile*, which means they can stretch and move to a limit without breaking; the process of hot-working mentioned above makes sure this characteristic is retained. A thick plate of steel, say 100mm thick, can be rolled into a very thin sheet of 0.1mm thickness without splitting (like my pastry normally does). The matrix of crystals and the bonds between them can be softened, reshaped and moved around.

Another property the bonds give metals is *elasticity*. If a metal is pulled or squashed by a force (within a certain range), it adjusts back to its original shape when the force is removed. It's similar to when a stretched rubber band is released and returns to its normal size and shape – unless it's overstretched, in which case it deforms. The same thing can happen to metals.

In combination, these characteristics – the bonds, ductility, elasticity and malleability – make metals resistant to cracking. This gives them a very special property that makes them ideal for construction: they are good in tension. It was this property of metals that revolutionised the way we build. Before, structures had been designed mainly for compression, but now for the first time, we could create structures that could stand up to significant compression *and* tension.

While pure iron is good in tension, it's too soft to resist the immense loads in larger structures because the bond between its crystals is quite fluid and flexes. So engineers of the past could make decorative pillars, but pure iron was not strong enough for large, complex structures. It needed to be strengthened somehow. The crystals that make up iron are arranged in a lattice, so scientists and engineers began devising ways to stiffen it.

One way to do this is to add atoms to the lattice. A simple (and tasty) illustration of this involves taking lots of Maltesers and rolling them under your hand on a table, during which you'll find that they move around very easily. But if you then add a few chocolate-covered raisins to the mix, you won't be able to roll them as easily as before. Okay, you can eat the experiment now, but the point is that the 'impurities' – the raisins – lodge themselves in awkward positions and stop the Maltesers from moving around as smoothly. Similarly, if carbon atoms are added to iron they jam the crystal lattice.

There is a balance. Too few carbon atoms and the iron is still too soft. Too many, and the lattice becomes so stiff that it loses its fluidity and the material ends up very brittle, cracking easily. As if this wasn't complicated enough, iron naturally contains some carbon (and other elements like silicon) as an impurity – usually too much – but the amount varies, so the quality of the iron varies. Scientists had great difficulty trying to determine precisely how much carbon to remove to create iron that was neither too soft nor too brittle. Results of their experiments include cast iron (which, being resistant to wear, is good for cooking pots, but is not used much in buildings because it's brittle, like an Italian *biscotti*); wrought iron (which is not used much commercially any more, and which has a texture more like the soft, luxurious chocolate-chip cookies I used to eat as a child in America); and steel. While wrought iron was a decent enough building material – the Eiffel Tower is made from it – steel turned out to be the ideal compromise between strength and ductility. Steel is simply iron with about 0.2 per cent carbon content. The process of removing all but 0.2 per cent of the carbon was originally very expensive, so until someone worked out how to manufacture steel cheaply and on a large scale, it didn't make a splash in the structural world. Engineer Henry Bessemer finally solved this long-standing problem and revolutionised the steel-making process, facilitating the development of railways across the world and allowing us to begin building skywards.

*

Henry Bessemer's father, Anthony, ran a factory that manufactured typefaces for the printing press that he kept under lock and key. The protection was designed to safeguard his secrets from his competitors, but the young Henry often broke in to try and figure them out. Realising that his disobedient son was adamant about learning a trade, Anthony relented, and began training him in the factory. In 1828, when he was fifteen, Henry left school to work with his father. He loved it: he excelled at metalwork, had a natural talent for drawing and eventually began making his own inventions.

During the Crimean War (1853–1856), Henry Bessemer turned his attention to the guns the French and British were using against the Russians. The principal drawback of these guns was that they could only fire one shot before they had to be reloaded. An elongated shell that could carry more explosive seemed like a valuable improvement, so Henry tested this in the garden of his home in Highgate, North London (much to the annoyance of his neighbours). The British War Office, however, wasn't interested in his design, so he showed it to the French emperor, Napoleon Bonaparte, and his officers.

Although impressed by the shells, the officers pointed out that the extra firepower would make their brittle cast-iron guns explode. As far as they were concerned, the shells were too big. Bessemer disagreed: the problem was the guns, not the shells – so he took on the challenge of finding a better way to make them.

He decided to improve the quality of the iron being used to make the guns by developing another way of casting it. He set about formally experimenting in his homemade furnace, but the invention that made his name happened almost by mistake.

One day, in his workshop, Bessemer was heating pieces of iron in a furnace. Even though he turned up the heat, a few pieces on the top shelf refused to melt. Frustrated, he tried

MANUFACTURE OF STEEL: THE BESSEMER PROCESS.

The Bessemer Process, developed for producing steel on an economic scale, led to radical developments in the construction industry.

blowing hot air into the top of the furnace, and then prodded the pieces with a bar to see if they had finally melted. To his surprise, they were not brittle like cast iron but instead were ductile and flexible. Noticing that they were the ones closest to the hot air, Bessemer realised that the oxygen in the air must have reacted with the carbon and other impurities in the iron – and removed most of them.

Until now, everyone had tried to purify iron by heating it with coal or other fuels in an open furnace. Bessemer decided to use a closed furnace with a current of warm air running through it – and without using any fuel. This is like blowing hot air into a pan which has a lid covering it, rather than heating up an open pan on a gas hob. You would normally expect burning gas to create more heat than hot air, but this is not what happened.

Bessemer must have watched cautiously as sparks emerged from the top of the furnace when the chemical reaction began. Then, a raging inferno started up – there were mild explosions and molten metal splashed around, erupting from the furnace. He couldn't even approach the machine to switch it off. Ten terrifying minutes later the explosions petered out. Bessemer discovered that what was left in the furnace was purified iron.

The furnace inferno was the result of an *exothermic* reaction: a chemical reaction that releases energy – usually in the form of heat – during the oxidation of impurities. After the silicon impurities had been quietly consumed, the oxygen in the air current reacted with the carbon in the iron, releasing a huge amount of heat. This heat raised the temperature of the iron far beyond what a coal-fired furnace was then capable of, so Bessemer didn't need to use external sources of heat. The hotter the iron became, the more impurities burned off, which made the iron hotter still, so it burned off even more impurities. This positive loop created pure, molten iron.

Now having pure iron to work with, Bessemer found it easy to add back precisely the right amount of carbon to create steel. Until his invention, steel's prohibitive manufacturing costs meant it was used to make cutlery, hand tools and springs, but nothing larger. Bessemer had just swept away that huge barrier.

He presented his work at the British Association meeting in Cheltenham in 1856. There was huge excitement about his process because his steel was almost six times cheaper than anything else available at the time. Bessemer received tens of thousands of pounds from factories all over the country to replicate his process. But his lack of understanding of chemistry was nearly the end of him.

When other manufacturers tried to reproduce Henry's methods, they failed. Furious at the amount they had spent on the licence to use the process, they sued Bessemer, and he returned all their money. He then spent the next two years trying to figure out why the process worked perfectly in his brick-lined furnace but not in others. Finally, he cracked it: the iron he was using contained only a small amount of phosphorus as an impurity. His peers, however, had been using high-phosphorus iron which, it appeared, didn't work in a brick kiln. So Bessemer experimented with changing the furnace lining, and realised that replacing brick with lime was the answer.

However, the perplexing and financially frustrating failure of his original process had bred a mistrust of Bessemer that meant no one believed him. Finally, he decided to open his own factory in Sheffield to mass-produce steel. Although it took a few years before suspicions faded, after that factories started manufacturing steel on a truly industrial scale. By 1870, fifteen companies were producing 200,000 tonnes of steel each year. When Bessemer died in 1898, 12 million tonnes of steel were being produced worldwide.

High-quality steel transformed the railway networks because it could be made into rails quickly and cheaply, and they lasted ten times longer than iron rails. As a result, trains could be bigger, heavier and faster, clearing up the clogged veins of transport. And because steel was cheaper, it could now be used in bridges and buildings – ultimately opening up the sky.

*

Without Bessemer's steel, I wouldn't have been able to design the Northumbria University Footbridge, which literally hangs on steel's ability to carry tension. The bridge was, in fact, the very first structure I worked on, fresh out of university. I can still vividly remember the first day of my brand new job, taking a packed Tube train to Chancery Lane in London, and being swept up and out of the station by the hurrying throngs of other professionals in suits. Feeling excited, nervous and a little awkwardly formal, I threaded my way along the pavement towards my destination – a five-storey office building clad in white stone.

My new boss was John, a slim man of average height, with straight, short dark hair, rimless glasses and a passionate love of cricket (something that, even though I grew up in India, I couldn't match). We went through some forms, a process made lively by his occasional ironic and funny observations; meanwhile I kept quiet about the fact that it was my 22nd birthday. Then he showed me his hand-drawn sketch of a new footbridge, made from steel, that was due to be built in Newcastle. The confident pencil marks showed that at the east end of the bridge a tall tower would support three pairs of cables. The cables in turn would hold up the main deck of the bridge. To counter-balance the weight of the bridge on this tower, a further set of cables would anchor it from behind. As I sat with John, looking at the drawings in front of me, I did a little dance inside. As far as I was concerned, this was as good a birthday present as a girl could get. I was thrilled that my first project was going to be this elegant and distinctive structure. Apart from its lovely aesthetic, however, this bridge had other nuances that made it, to my eyes, even more beautiful.

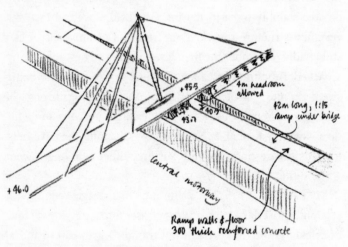

A working sketch of the Northumbria University Footbridge by John Parker.

The bridge is a 'cable-stayed' bridge, one famous example of which is the Millau Viaduct in France. Its gently curving deck is held in place by seven pillars, from which cables fan out in the shape of a sail, giving the impression that the bridge is floating 270 m above the Tarn valley. Cable-stayed bridges have one or more tall towers to which cables are attached; the deck is pulled down by gravity, and is held in place by cables, which are always in tension. The tension forces are channelled through the cables directly into the tower. The tower in turn compresses and the forces flow down into the foundations on which the tower is supported; the foundations spread the forces out into the ground.

The Millau Viaduct in France is an elegant example of a cable-stayed bridge.

As a fresh-faced engineer, designing the cables for the Northumbria Footbridge (which were as thick as my fist) was a real challenge. If you take a metal ruler, representing the steel deck, and use three pairs of rubber bands to mimic the cables, you'll find that you have to pull on each band just the right amount before they're all taut and supporting the ruler evenly so it lies flat. If you pull too hard on the three bands on one side, the ruler tips over sideways. If you pull too tightly on the central pair, it bows upwards. Now imagine the same effect, but on a real, full-sized bridge.

I used software to create a three-dimensional computer model to recreate the bridge beams that run under the deck and the cables that run from the deck to the mast. Then I simulated gravity on the structure. I also had to consider the weight of all the people that would stand on the bridge, and that they might congregate on different parts of the bridge at different times. For example, during the Great North Run, in which athletes run along the motorway below the structure, cheering crowds might stand on one side as the runners approach them, then walk to the other side to watch them continue into the distance. I had to think about 'patterned loading' – I modelled people standing on the bridge in different configurations. No matter where people stood, the cables had to remain tight to support the deck. If the cables were not in tension they would become floppy, and the deck would lose its support. To stop this from happening, I added extra tension to the cables artificially.

Cables can be tightened up using a jack – which is a tube with clasps on each side. Each cable had at least one break in it where a jack could be installed. The clasps each held a bit of cable either side of the break. The jack can be adjusted to pull the ends closer together (to tighten the cable) or further apart (to loosen it), therefore altering the amount of force in the cable. If you look at the cables fanning out from the tower of my footbridge you'll see that they have connector pieces – where the cables look briefly thicker than the rest of their length: those are the points at which the jacks were temporarily connected. This is like replacing the rubber-band cables in our demonstration with shorter ones, but then stretching them out to the same length as before. This puts more stretch in the rubber bands – they contain a higher pulling or tension force.

The key to building a cable-stayed bridge is balance. If you use a thin piece of card as a deck and pull on the rubber bands, the card simply lifts up. If you replace the thin card with a book, then you can pull on the bands to make them taut without deforming the book. Once the stiffness and weight of the deck and the tension in the cables are reconciled and calibrated, you can then work out what the force is in the cables. When I did the drawings of the bridge, I added notes stating how much each cable needed to be tightened to stop it going slack.

The engineer's job is a lot like plate-spinning. You have to plan for, and control, a multitude of problems simultaneously. Take temperature: like all structures, my bridge is affected by it. Throughout the year, to varying degrees (depending on the season), it will be heated and cooled. Steel has a 'coefficient of thermal expansion' of 12×10^{-6}. This means that for every 1 degree of change in temperature on a piece of material 1mm long, the material will expand or contract by 0.000012mm. This may sound small, but my bridge was about 40m long and had to be designed for a temperature range of 40 degrees. The savvy among you will argue that the British summer is not 40 degrees warmer than the winter, and you would be correct, but the steel itself will get much hotter than the air as it absorbs heat from the sun. We're looking at the range of temperature experienced by the steel, not the air, in the most extreme (but reasonable) weather we can anticipate.

This adds up to an expansion of nearly 20mm. If I fixed the ends of the bridge to stop it from expanding or contracting, a large compression force would build up in the steel deck when it got warmer, and a large tension force would build up when it cooled down. The problem is that this expansion and contraction could happen thousands of times over the life of the bridge; this constant pulling and pushing can gradually damage not only the steel deck itself, but also the supports at either end.

To prevent this, I allowed one end to move. (In larger bridges, or bridges with many supports, you can create 'movement joints' in multiple places. You can sometimes feel your car 'boing' as you drive over them.) Because the movement on this bridge was relatively small, I used a 'rubber bearing' to absorb it. The steel beams which made up the deck were supported on these bearings, which were about 400mm wide, 300mm long and 60mm thick. When the steel expands or contracts, the bearings flex, letting the bridge move.

I also needed to think about vibration and resonance. I've already explored how an earthquake can cause a building to resonate, just as an opera singer can shatter a wine glass when she hits the right note. With the footbridge, I was concerned about whether resonance could make pedestrians feel uncomfortable. Heavy bridges, like those made from concrete, generally don't suffer from this problem because their weight stops them from vibrating easily. But the steel deck was light, and its natural frequency was close to the frequency of walking pedestrians, which meant it was in danger of resonating. So we connected tuned mass dampers with strong springs to the underside of the deck. These work in a similar way to the giant pendulum inside the Taipei tower, absorbing the sway and stopping the deck from vibrating too much. You can't see these tuned mass dampers unless you look carefully at the bottom of the deck from the road underneath the bridge (perhaps while stretching your legs on the Great North Run). If you do, you'll notice three steel box-like objects hidden between the bright-blue-painted beams.

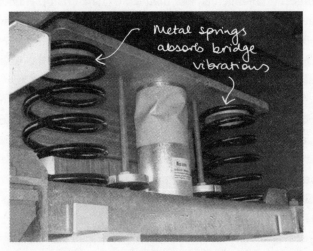

A type of tuned-mass damper, similar to those used on the Northumbria University Footbridge.

Once I was sure that my bridge was stable in its final configuration, I had to work out exactly how it would be built. As it was too large to be transported to Newcastle fully constructed, I went to a steel fabricators' factory in Darlington. Amid showers of sparks cascading from a welder's arc, we discussed some options. We would have to bring the bridge to the site in pieces that fit on the back of lorries, so we looked at splitting it in various places, checking how those sections could be installed and supported safely until the cables had been tied in; like a sculpture that would need to support itself even while each piece was being placed.

We also had to consider how to cause minimal disruption to the public. Since the structure was to span a motorway, we decided the best approach was to bring it to site in four pieces, connect them together, and then use a crane to lift the assembled bridge into place. A one-of-a-kind monster crane was booked to do the job.

Months of planning went into ensuring that the bridge was hitched up without a hitch. First, the crane itself arrived in pieces at the start of a bank-holiday weekend, and roads were closed off as it was assembled by swarms of steel fixers. Meanwhile, the four steel sections of the bridge were transported from Darlington to a nearby car park, where they were joined together, like a jigsaw puzzle, to make the deck.

The plan was to hoist the steel deck into place, and then to attach the cables. I had designed the deck such that it needed all three sets of cables to resist both its own weight and the weight of pedestrians crowded on top of it. This meant that, until the cables were in place, it needed extra support on site, so I had also calculated that the deck could stand up with a single support at its centre (it had less load on it in this configuration as the public wouldn't have access). We erected a temporary steel column in the central reservation of the motorway.

The motorway was closed. The crane swung into action. The deck was lifted up from the car park and lowered into place, its ends held up by their permanent concrete supports, and its centre by the temporary steel one. The deck was disengaged from the crane, and the motorway reopened. This complex operation took just three days.

Over the next few weeks the rest of the bridge was assembled. The mast was lifted into place using a crane and then anchored to its concrete base with bolts. The all-important cables could then be installed in pairs starting from one end of the bridge. Every time a new pair of cables was connected, the tension was adjusted using a jack. Once the cables were all in and adjusted one final time, the road was closed again, the temporary steel column removed, and the bridge was complete.

I'm not normally excited about getting up early, but my eyes were already wide open at 5am on the day I travelled to Newcastle to visit my completed bridge, which was now ready and open to the public. After taking a first small step, which felt to me like a giant leap, I walked back and forth across the bridge a number of times. I skipped and I ran. The solid steel beams, the taut cables, the rubber bearings, the tuned mass dampers – they all reminded me of the time, only a few months ago, when I had painstakingly designed them. Details that perhaps no one would notice except me – but they made me happy.

At one end of the bridge there was a bench. I sat there, grinning, for a while, watching bleary-eyed students walking across the deck from one lecture to another, all of them oblivious to the pleasure it gave me to experience my first physical contribution to the world.

ROCK

I've been known to stroke concrete. Others might feel the irresistible urge to pat a little kitten or handle an object in a museum, but for me it's concrete. It doesn't matter if it's a smooth, stark grey surface, or one with little stones visible, or even one left intentionally rough – I have to know what the texture feels like, how cold or warm it is. So you can imagine how I felt when I visited Rome and saw tonnes of ancient concrete above my head, but too far away to reach.

The Pantheon in the Piazza della Rotonda in Rome is one of my favourite structures. Built by the emperor Hadrian around AD 122 (at about the same time as he was building a wall to divide England from Scotland), it has stood strong ever since in a variety of guises – temple to the Roman gods, Christian church, tomb – though barbarians removed what they could and Pope Urban VIII even melted the ceiling panels to make cannons. A triangular pediment supported by a portico of sixteen Corinthian columns greets you at its entrance. Inside, the rotunda is topped with a dome punctuated by a circular opening (*oculus* – Latin for eye) through which streams an almost otherworldly shaft of light. It's an atmospheric and beautifully proportioned building. I'm overwhelmed by its sheer scale when I wander around in it, bumping into people as I stare up at the beautiful roof. Even now, it's the largest unreinforced concrete dome in the world. The Romans really honed their craft, creating an engineering masterpiece from a revolutionary material they called *opus caementicium*.

The giant concrete dome and oculus at the Pantheon in Rome, Italy.

For me, what's special about concrete is that its form is indeterminate: it can be anything. It starts as rock, then becomes a lumpy grey liquid that can be poured into a mould of any shape and left while chemistry takes over, turning the liquid back into rock. The end product could be a circular column, a rectangular beam, a trapezoidal foundation, a thin curvy roof, a giant dome. Its amazing flexibility means it can be formed into any shape; because of its huge strength, and because it lasts an extremely long time, concrete is, after water, the most-used material on the planet.

If you crush most types of rock into a powder and add water, you end up with an uninteresting sludge, the two parts don't hold together. But something strange happens when you heat certain rocks up to really high temperatures. Take a mixture of limestone and clay, for example, and fire them in a kiln at about 1,450 degrees Celsius, and they will fuse into small lumps without melting. Grind these lumps into a very fine powder and you've got the first ingredient of an incredible material.

The powder is called cement. It's a dull grey colour and might not look particularly impressive. But because it's been burned at very high temperatures, the parent materials are chemically changed. If you add water to this powder it doesn't turn into a sludge – instead, a reaction called *hydration* begins. The water reacts with the calcium and silicate molecules in the lime and clay to create crystal-like rods or fibres. These fibres give the material a jelly-like structure – a matrix – which is soft but stable. As the reaction continues, the fibres grow, and they bond to each other. The mixture becomes thicker and thicker until, ultimately, it solidifies.

So water + cement powder = cement paste. Cement paste hardens into a rock incredibly well, but it has its drawbacks. For a start, making it is expensive. The process also uses a large amount of energy. And importantly, hydration releases lots of heat. Once the chemical process finishes, the cement cools down, and as it cools it shrinks. And cracks.

Fortunately, engineers realised that cement paste binds solidly to other rocks, and began adding *aggregates* (small, irregular pieces of stone and sand of varying sizes) to the mixture. The aggregates help to reduce not only the amount of cement powder being used (and hence the amount of heat being released), but also the energy consumption and hence cost. The cement paste undergoes the same chemical reaction, creating fibres that in turn bind strongly to other fibres *and* the aggregates – and the whole mass solidifies to give us the concrete we are familiar with today. So water + cement powder + aggregate = concrete.

To make good concrete, the proportions of this mix need to be right: too much water, and not all of it will react with the cement powder – and the concrete will be weak. Too little water and all the powder doesn't react and, again, the concrete ends up weak. For the best result, *all* the water needs to react with *all* the cement powder. And the mixing itself needs to be right too: concrete can end up poorly if it isn't stirred properly. The larger, heavier stone aggregates settle to the bottom, leaving the fine sand and cement paste at the top, making the concrete inconsistent and weak. That's why concrete trucks have giant rotating drums – the mixture is continually sloshed around so that the aggregates are nicely distributed throughout.

Ancient engineers didn't have such trucks, but their formula for concrete was pretty similar to ours. They too burned limestone, and powdered it then added water to create a paste with which to bind stones, bricks and broken tiles. However, their mixture was much lumpier and thicker than ours is today. But then the Romans found something even better. In the land around Mount Vesuvius was an ash they called *pozzolana*. Instead of using burnt limestone as a cement, they tried this ready-made ash. When they mixed it with lime, rubble and water, their resulting concrete hardened as they expected. But this mixture also hardened underwater. That's because the *pozzolanic* chemical reaction did not need carbon dioxide from the air to help it along: the mixture could harden without it.

To begin with, the Romans didn't appreciate the amazing potential of the material they had made, and they only used it in small structures in a tentative way. They used it to strengthen the walls of their homes and monuments – sandwiching a layer of concrete between two layers of brick. After all, how did they know that it wouldn't crack and crumble in a few years like plaster did? As the years passed, of course, they realised that this incredibly resilient substance was nothing like plaster, and concrete became a commonly used material. And because it solidified underwater, they could build concrete foundations for bridges in rivers, solving the problem they'd had so far in trying to cross vast stretches of water.

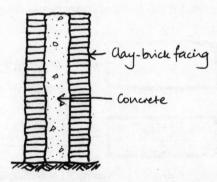

A Roman concrete sandwich. In Roman construction, the concrete wall was faced with a brick layer on both sides.

The Romans frequently used arches in their constructions, and concrete is a good material for arches. For one thing, it is incredibly strong. If a standard brick made from fired clay can carry the weight of five elephants, a similar brick made from relatively weak concrete can carry fifteen. In fact, a brick made from one of the stronger mixes of concrete can carry 80 elephants. And its strength can be changed, depending on the exact proportion of ingredients you add to the mix. Unlike bricks and mortar – where mortar is usually weaker than brick and more susceptible to crushing – concrete is cast monolithically (in large continuous chunks) and doesn't have weak links in the same way: its strength is maintained uniformly across its whole body. Ultimately, of course, if the compression load is large enough, concrete will crush and crumble, but it takes a lot of load (or a good number of elephants) to get to this point.

Concrete is, however, a fussy material. It loves compression, and for millennia it was used this way, being squashed in foundations or walls. But it dislikes being pulled apart. Its resistance to tension is minimal; in fact it cracks if tested at loads less than one-tenth of what it can resist in compression. This is another

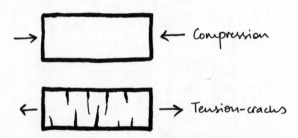

Fussy concrete prefers to be in compression. At even relatively low loads of tension, concrete will crack.

reason why the Pantheon impresses me so much. The Romans really understood how concrete works, and how domes work, and even though concrete wasn't the ideal material to use to build this immense structure, they still used it – and used it well.

To understand why making a dome from concrete is challenging, start by making an arch. If you bend a long, thin rectangular strip of card and place it on a table, you'll find it won't hold that curve on its own. It simply collapses. To make your arch stand up, position an eraser on the table against each of the outer edges of the curved card. The ends of the original, unsupported card arch pushed outwards, collapsing the structure; this time, however, although the arch still pushes outwards, the sideways friction between the eraser and the table reacts against the push from the base of the arch. This is Newton's third law of motion: every action has an equal and opposite reaction. The base of the arch exerts a pushing 'action' on its support – and its support keeps it stable by 'reacting' against this force.

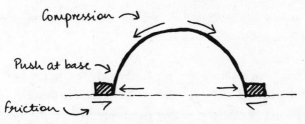

Forces flow around an arch, and then push out at the base.

Domes are similar to arches, but in three dimensions; the third dimension adds a layer of complexity. If instead of having one card you cut many long thin strips, then stack them one on top of the other and stuck a pin through the centre of the stack, you could still curve them downwards to create an arch. But you could also fan them out through 360 degrees (so that they form lines a bit like the longitudes of the Earth), thereby creating the shape of a hemisphere or dome. This dome, though, will be no more stable than your original, unanchored arch: it won't retain that hemispherical shape on its own. To hold it in place you could arrange a ring of erasers on the table, one at the base of each strip. Or you could try something smarter, such as using rubber bands, arranged like the latitudes of the Earth, to tie the dome together. With rubber bands in place, you can remove the erasers and the dome still stands.

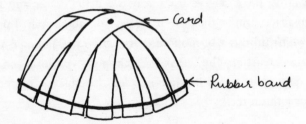

When 'tied' sufficiently, the forces that flow around a dome do not push out at the base.

What this means is that the supports for a dome do not feel any horizontal pushing force on them (unlike the arch). But you'll notice that the rubber bands are in tension: they are stretched and resist the push of the card strips. So yes, each of the strips is in compression individually along their 'longitudes', but you need tension to hold the strips together in the 'latitudes'.

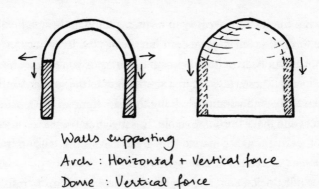

Walls supporting
Arch : Horizontal + Vertical force
Dome : Vertical force

The difference between where forces flow in an arch as opposed to a dome.

Viewed from the piazza, the Pantheon looks quite shallow, but in fact the inside is almost perfectly hemispherical. It appears shallow from the outside because the base is much thicker than the crown: the concrete at the top of the dome is only 1.2m thick, but by the time it reaches the base it has increased to more than 6m. Making it thicker towards the base meant the dome could resist higher tension forces – more material, more resistance.

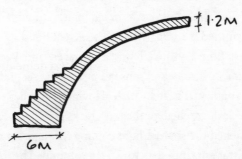

The widening stepped rings around its base act to strengthen the dome of the Pantheon.

But the Romans went even further, adding more stability in the form of seven stepped concentric rings (which you can see from the outside, a little below the oculus, if you're somehow airborne). These rings act in a similar way to the rubber bands from our demonstration, helping to resist some of the tension forces and make the dome stable. This ingenious design ensured that, even though concrete isn't great at resisting tension, the Romans succeeded in making it work.

While thicker concrete might solve some problems in resisting tension, it also creates problems of its own. The thicker the dome, the more the cement content – which means it generates more heat, and the more it shrinks when it cools. As it shrinks, it pulls itself apart and, since concrete can't resist this tension, it cracks. The Romans were worried that the base of the Pantheon's dome would suffer extensive cracking. It's believed that the series of squares which are inset all around the inside of the dome, which are part of its unique visual aesthetic, are there to allow the concrete to cool down more quickly and evenly, minimising cracking. Even so, engineers studying the Pantheon have found cracks in the base of the dome (ancient ones that occurred while it was being built) – though they haven't undermined the integrity of this ancient building.

The first time I visited I was a teenager, and I loved this building for its beauty and sense of peace. The second time, as a trained engineer, I gazed – no less lovingly – at the recesses in its surface and searched for the fine cracks at its base. For a long time I watched the shaft of light coming through the oculus at the top of this amazing structure. I left astounded by the dome's scale and apparent simplicity of form, but conscious of how complex it must have been to construct it so many years ago.

I often wonder whether, like the Pantheon, the structures we design and build today will still be around, and in such good condition, in 2,000 years. It seems inconceivable.

*

After the collapse of the Roman Empire in the fifth century, the Dark Ages – or as I like to call them, the Crumbly Ages – began, as the Roman recipe for concrete was lost for almost 1,000 years. We reverted to a move primitive way of life and concrete only re-emerged in the 1300s. Even then, engineers continued to struggle with the fundamental problem of concrete cracking in tension. It was only centuries later that the true magic of concrete was discovered, by an unlikely hero, in the most unexpected of places.

In the 1860s, French gardener Joseph Monier became fed up with the fact that his clay pots would constantly crack. He tried making pots out of concrete instead but found that they fractured just as much. Randomly, he decided to reinforce the concrete by embedding a grillage of metal wires within it. This experiment could have failed for two key reasons: first, the concrete might not have actually bound to the metal reinforcement (there was no reason to think that it would), so the metal would only create more weak points in the pot. Second, during the change in seasons, the metal and concrete would expand and contract at different rates, creating yet more fissures. Unwittingly, Monier created a revolutionary pot that remained solid and barely cracked.

Like most metals, iron and steel (as we've seen) are elastic and ductile, and they're good in tension: they don't crack when pulled. Metals aren't brittle like brick or concrete. So by combining concrete (which breaks in tension) with iron (which can absorb tension loads), Monier had created a perfect marriage of materials. In fact, an ancient version of this principle can be found in

Morocco, where the walls of some Berber cities were made of mud with straw mixed in: a mixture known as *adobe*, also used by the Egyptians, Babylonians and Native Americans, among others. Straw fulfils a similar function to metal in concrete; it binds mud and plaster together and stops it from cracking too much because the straw resists tension forces. The plaster on the walls of my Victorian flat has horse hair mixed into it for the same reason.

Having exhibited his new material at the Paris Exposition in 1867, Monier then expanded its application to pipes and beams. Civil engineer Gustav Adolf Wayss from Germany saw the material and had visions of building entire structures with it. After buying the rights to use Monier's patent in 1879, he conducted research into concrete's use as a building material, and went on to build pioneering reinforced concrete buildings and bridges across Europe.

The marriage of steel (which replaced iron once the use of the Bessemer process spread) and concrete appears so obvious today that it seems almost inconceivable to me that the two weren't always used together in this way. In every concrete structure I design, I use steel *reinforcement bars* – long, textured rods between 8mm and 40mm in diameter that are bent

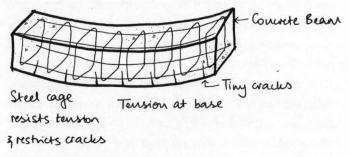

A perfect marriage of construction materials: a steel cage provides reinforcement for concrete, resisting tension and restricting cracking.

into different shapes and tied together to form a grid or mesh to bind the concrete. My calculations tell me where the concrete will be in tension and where it will be in compression, and I distribute steel bars within it accordingly.

Contractors take my drawings and set the dimensions and shapes of every single steel bar in the project, and calculate their weight. These schedules are sent to a factory, and a few weeks later real bars appear, which are fixed into shape before the concrete is poured around them.

As the chemical reaction in the concrete mixture progresses, steel and concrete form a strong bond. Just as cement paste binds strongly to aggregates in the mix, it also sticks to the steel. And once intertwined, steel and concrete are very difficult to separate. They have near-identical thermal coefficients – which is to say that they expand and contract by almost identical amounts under the same changes in temperature. When a concrete beam bends under gravity and is squashed at the top but pulled apart at the bottom, the concrete cracks at the bottom. These cracks are fractions of a millimetre wide and often not visible to the human eye – but they are there. Once this happens, the steel bars in the base of the beam are activated, and resist the tension loads keeping the beam stable.

Steel reinforcement is now part of the DNA of how we build modern structures. Many construction sites around London have small windows in the protective hoardings that surround them. As you can imagine, whenever I walk past one I can't resist taking a peek, curious to see what's going on inside. No matter what the site, I always see big piles of steel reinforcement bars ready to be tied together, or steel cages already made up inside wooden moulds. When the trucks with rotating drums appear, they pour a thick stream of concrete into the moulds, after which workers use short poles attached to a power supply to vibrate the concrete, mixing it to make sure that the different-sized aggregates are well-distributed throughout. Engineers like me have made sure that the gap between the steel bars is big enough to allow the concrete to flow easily around them. As a young engineer, my first boss John told me, 'If a canary can fly out of your steel cage, the bars are too far apart. If it suffocates, they're too close together.' It's a lesson I've never forgotten. (No canaries were harmed in this thought experiment.)

Once all the concrete has been poured and mixed thoroughly, the workers flatten the top of it with huge rakes and leave it to solidify. But this incredible material has one more secret in store. Over the next few weeks, the bulk of the chemical reaction will finish, it's tested, and results show that it has reached its target strength. In fact, its strength continues to grow – very slowly – over months, and even years, plateauing to a steady number far into the future.

*

Nowadays, we use concrete for many structures, creating skyscrapers, apartment blocks, tunnels, mines, roads, dams and countless others. In ancient times, different civilisations employed different materials and techniques that were suited to their indigenous skills, climate and surroundings. Today, concrete is universal.

Scientists and engineers are constantly innovating, trying to make concrete even stronger and longer-lasting than it already is. One recent invention has been 'self-healing' concrete, which contains tiny capsules with calcium lactate. These are mixed with the liquid concrete, but the capsules have a fascinating secret. Inside is a type of bacteria (normally found in highly alkaline lakes near volcanoes) that can survive without oxygen or food for 50 years. The concrete, mixed with these bacteria-filled capsules, hardens. If cracks form in the material and water seeps in, the water activates the capsules, releasing the bacteria. Habituated as they are to alkaline environments, these escapees don't die when they encounter the highly alkaline concrete. Instead, they feed on the capsules, combining the calcium with oxygen and carbon dioxide to form calcite, essentially pure limestone. With calcite filling the cracks in the concrete, the structure repairs itself.

There are other challenges. Five per cent of human-created carbon dioxide comes from making concrete. Using concrete in small amounts is not particularly unfriendly to the environment, but we use so much of it that the emissions quickly add up. Some of the CO_2 comes from the firing of limestone to create the cement, but the rest comes from the hydration reaction. The amount of cement being used in the mix can be reduced by replacing a proportion of the cement with suitable waste materials from other industrial processes, such as 'ground granulated blast furnace slag' (GGBS), which is created during the manufacture of steel. Using these waste materials doesn't affect concrete's strength too much but can save tonnes of carbon. You can't use them for all types of construction, because these ingredients have other effects on the mix. They can make the concrete take longer to solidify, or make it stickier, and hence harder to pump up many storeys, which is definitely a challenge when constructing skyscrapers.

'My' skyscraper, The Shard, uses concrete and steel in a really clever way that neatly reconciles the different requirements of office and residential areas. In typical office buildings, the aim is to create large, open spaces with few columns. Steel is often the material of choice because it behaves well in both tension and compression meaning that steel beams can span further than concrete ones of the same depth. Moreover, compared to apartment buildings, offices need a lot of air-conditioning machines, ducts, water pipes and cables. The I-shaped construction of steel beams, and the regular gaps between adjacent beams, leave plenty of space to hide these away. Steel structures are also lighter than their concrete equivalents, so the foundations can be smaller as well.

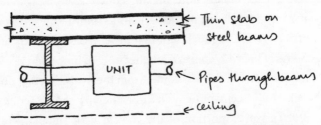

Arranging the steel beams and concrete floors for an office building.

On the other hand, residential buildings and hotels have floors that are sub-divided into flats and rooms, so you're not under as much pressure to create huge open spaces. You can hide concrete columns in walls to support flat concrete slabs. Concrete floors are thinner than steel ones, so you can fit more storeys into a concrete building of the same height. There are fewer cables and smaller ducts to run, and these can be attached to the bottom of the slabs. Concrete also absorbs sound better, so you get less noise transfer between floors – this doesn't matter so much in an office where you, hopefully, don't sleep.

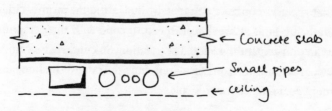

Arranging the concrete floors for a residential building.

Since The Shard has offices on its lower storeys and a hotel and apartments on its higher ones, we used different materials in different places. The lower storeys are made from steel columns and beams to create space in the offices; the higher storeys from concrete to create privacy. While it may seem obvious to use the right material in the right place, it's actually quite an unusual thing to do, and only a handful of structures globally have so far adopted this design. One possible reason is that it's arguably logistically easier (and possibly cheaper) to use the same material throughout, but I'd counter that by saying you achieve a better design for the long term, and it's more sustainable because you use less material. Another reason is that multi-use buildings simply aren't as widespread as single-use buildings. But with the construction of more and more multi-use buildings, I expect the multi-material method will become more common.

Using the materials we have in an efficient way is good engineering. We often think of concrete as being old-fashioned because of its ancient roots, but it's still very much part of the future too. Scientists and engineers are working on new super-strong mixes, and trying to figure out how to make concrete more eco-friendly. Perhaps one day we may find a new material that replaces concrete completely. But in the meantime, cities are being built at breakneck speed to cope with the demands of an ever-expanding, global population. So concrete buildings will grace our horizons for a long time to come. Which means more concrete for me to stroke.

SKY

Over the years I've worked on a range of projects, from the steel footbridge in Newcastle and concrete apartment blocks in London, to the refurbishment of the brick railway station at Crystal Palace. But skyscrapers have become one of my specialities – which is ironic, because I have no head for heights.

Don't get me wrong: I won't freeze up and go bulgy-eyed, like James Stewart at the beginning of *Vertigo*. I don't collapse into a blubbering mess when I look down from a great height, even if my legs have turned to jelly. But there's no doubt that it makes for some uncomfortable moments at work. Most days, I'm safely sat at a desk inside an office (reassuringly low down on the ninth floor). But sometimes I have to don the classic clobber of my profession – hard hat, hi-vis jacket, steel-toed boots – and climb up a structure I've been designing.

So it was with a mixture of excitement and anxiety that I got off the train at London Bridge in May 2012, took a right out of the station and walked up the street towards a plyboard door painted bright blue – a part of the site hoarding, ignored by the thousands of commuters on their way to work. This was once the entrance to The Shard: a sharp contrast to the gleaming glass and white steel construction that welcomes you today.

The Shard is now a landmark in London, England.

Moving past the plyboard portal, I entered a maze of plastic barriers and wove my way through, slightly worried that I'd get lost, as the fenced pathways were arranged differently from the last time I'd visited. Eventually I stepped tentatively into a cage-like elevator – a hoist – that was inclined slightly to match the angle of the tower. It shuddered and groaned then shot up rapidly, while my eyes stayed glued to the building, not daring to look down. (Knowing that The Shard's elevator was the first inclined hoist ever to be stuck to the outside of a tower was cool, but it did nothing to lessen my discomfort.) When the elevator finally ground to a halt, I emerged halfway up the building. It was quiet and deserted, and its skeleton was bare: rust-coloured steel columns towered above a firm, blotchy-grey concrete floor. Resisting the urge to stroke it, I tried to picture what this place might look like when it was full of people, furniture and activity. On that day, it was quiet.

I willed myself back into the hoist, this time rising to the highest level it accessed – the 69th floor. Here it felt completely different. The structure was open to the elements. Metal barriers protected the edge of the building, as the glass was not yet installed. The solitude of the lower levels was replaced by a flurry of activity – workers shouting instructions, pieces of steel clanging, cranes beeping as they lifted beams, and concrete spewing out of quaking pumps. Above me rose the crown of the tower – its elegant spire – which I had worked on. Another eighteen flights of stairs led to the highest floor. It suddenly hit me that this was the first time I'd been able to go there, as it hadn't been finished on my previous site visits. Today was truly special.

At the top step, though, I had to stop. The tapering shape of the tower meant that this level – the 87th floor – was relatively small. Even standing at the staircase, which ran through the centre of the floor, I seemed close to the edge. My stomach churned. I suppressed a rising feeling of fear. Fresh, chilly air entered my lungs as I took calming breaths with my eyes closed. When I felt less dizzy, I opened my eye (that's right, just the one).

I was at the intersection of the sky and humanity. After months of making models, doing calculations and creating drawings, I was finally seeing the project made real. It felt so much larger and more tangible than the sketches on a piece of paper or drawings on a computer screen. This phase of construction is a thrill: a moment when the niceties of false ceilings and floors are missing, there isn't the restriction of a facade, and the general public has never crossed the threshold. To me it felt like having a backstage pass for the rehearsal of a big rock concert – a privileged glimpse of all the stuff that will soon be hidden away and embellished, but which forms the backbone of what we will finally see. Visiting the site filled me with awe for the object we had created. It motivated and refreshed me, and reminded me why I love the creative process of design and construction, particularly for skyscrapers.

*

If you were to draw a graph of humanity's tallest buildings over time, which is exactly the sort of thing I might happily spend an evening doing, you would see that it suddenly shoots skywards around the 1880s. For millennia, the Great Pyramid of Giza (at 146m) held the record as the tallest human-made structure in the world. It wasn't until medieval times that this record was surpassed, by Lincoln Cathedral (160m), which held the title from 1311 until 1549, when a storm snapped its spire. This made St Mary's Church in Stralsund in Germany (151m) the tallest building – until it, too, lost its spire, to a lightning strike in 1647. It was replaced by Strasbourg Cathedral (which was a mere 142m, but by now the Great Pyramid had eroded so much it didn't reach 140m). The real quest for height began in the nineteenth century, when the first skyscraper was erected in Chicago in 1884. Admittedly, at 10 storeys – a mere 42m – it's hardly what we think of as a skyscraper today, but it was the first tall building to be supported by a metal frame. In 1889 the Eiffel Tower became the first building to hit the 300m mark. Since then our ambitions, and our buildings, have soared. It took nearly 4,000 years to beat the height of the pyramids – shaky spires notwithstanding. But in the past 150 years, our structures have grown from about 150m tall to over 1000m.

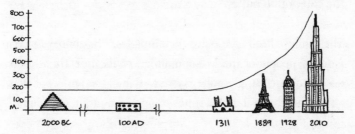

Plotting the heights of the tallest buildings over time demonstrates how technical innovations over the past century have accelerated how high we can build.

Isaac Newton famously said that 'If I have seen further, it is by standing on the shoulders of giants.' Standing at the top of the tallest tower in western Europe (310m), and aware of all the material and techniques that had gone into its making – the clanging steel and beeping cranes to name a couple – I was vividly reminded of how we got here, of the key people in our history who helped unlock the sky. Newton, of course, was one of them: without his Third Law of Motion, for example, I wouldn't be able to calculate the forces at work in an arch. But there are others who pushed us to think outside the box (of simple, single-storey dwellings) and who created the cranes and elevators without which we would still be stuck at ground level or thereabouts. The Shard is built not just on innovative foundations but on a legacy of historical ideas and advances that revolutionised construction and made our skyscrapers possible. For a start, to get a tall building off the ground we have to get things off the ground. Before cranes, the difficulty of this task seriously limited our construction ambitions – until, that is, Archimedes (287–212 BC) invented the compound pulley.

*

The pulley itself pre-dates Archimedes. In approximately 1500 BC people of the Mesopotamian civilisation (in what is now Iraq) used single-pulley systems to hoist water. A pulley is a suspended wheel with a rope wrapped around it. One end of

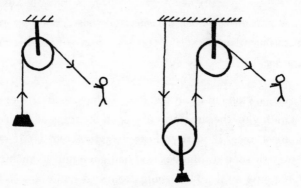

Simple (above left) and compound pulleys (above right).

the rope is tied to the heavy object that needs to be lifted – like a bucket – while a person pulls on the other end. It was a very practical tool, because you could lift objects while standing on the ground and pulling downwards, using gravity to help you. Until the pulley was invented, you had to find a level that was higher than your object's destination and pull upwards. The pulley changed the direction of the force, which meant we could move larger loads.

Archimedes, however, had a restless imagination that he applied to mathematics, physics and even weapon-making, as well as engineering. He improved the pulley by wrapping the rope around not one wheel but several. With one pulley, the force you have to exert to lift a load of a certain weight is equal to that weight. So a 10kg mass needs a force of 10kg x 9.8m/s^2 (the gravitational pull), which equals 98N. (The N stands for *newtons*: named after the scientist, and another reminder of how key a figure he is for engineering – without his Law of Universal Gravitation I wouldn't be able to make this calculation.) The amount of *energy* you expend is the force you've applied multiplied by the distance. With a single pulley, if you want to lift this weight by 1m, you have to pull the rope 1m as well, so the energy you've used is 98N x 1m = 98Nm (i.e. newton metre).

If you use two pulleys, however, while the energy you expend must remain the same (since you're moving a fixed weight by a fixed amount), you halve the force needed. The reason for this is that the weight is now supported by not one but two sections of rope. Each section of rope needs to move by 1m to lift the weight by 1m, which means you have to pull the rope by 2m. Since the energy is the same, but the distance is doubled, the force you apply is halved. The same principle applies for three pulleys, or ten.

Archimedes made a radical claim to his ruler, King Hiero II, that *any* weight could be moved using his compound pulley system. Unsurprisingly, Hiero was sceptical and demanded that Archimedes prove it. One of the largest cargo ships from the king's arsenal was heavily loaded with people and freight. Hauling it to the sea with ropes normally took the full strength of dozens of men, but Hiero challenged Archimedes to do it alone. Watched by the king and an assembled crowd, Archimedes set up an arrangement of pulleys, wrapped a rope around them, attached one end of the rope to the ship, and pulled on the other. According to *Plutarch's Lives* (biographies believed to have been written in the early second century) 'he drew the ship in a straight line, as smoothly and evenly as if she had been in the sea.'

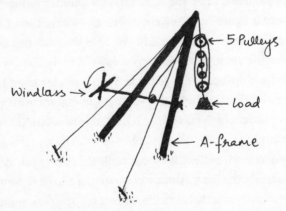

A Roman crane using a five-pulley system.

The Romans recognised the multiple pulley's potential and developed it further by incorporating it into their cranes. Two staves of wood arranged in an inverted V formed the crane's skeleton. The top ends of the staves were fixed together with an iron bracket and the base was anchored to the ground. Between these two staves a rod would be set horizontally (creating an A shape) to act as a windlass: i.e., a rope could be attached to it and then raised or lowered by rotating it, just like the apparatus used to operate a bucket in a well. Fixed to the top of the crane was a two-wheel pulley block; a rope was threaded from the windlass through this and into a third pulley positioned just above the load. At either end of the windlass were four handle-like spikes that could be used to turn it, thereby raising or lowering relatively big loads with ease. If the Romans had to lift something larger they added more pulleys and more rotating sections, and replaced the four turning spikes with a large wheel called a *treadwheel*.

Using a crane with pulleys, a Roman labourer could lift loads 60 times heavier than an ancient Egyptian could handle. And although they are much bigger, the cranes we use today still work on the same principle. Long, square hollow pieces of steel are assembled into a frame to form a very tall tower, and a long arm or *jib* is attached. The jib holds the all-important multiple pulley system, and the human muscles and spike-handles of the Roman version are replaced by petrol power. The jib moves right and left, through 360°, carrying multiple tonnes of steel or glass, attached safely to the modern version of Archimedes' invention.

*

By understanding the potential of cranes and arches, the Romans were able to build bigger. But their abilities were matched by their ambition: they were prepared to think bigger as well. As their empire grew, and the population along with it, the Romans found their towns expanding into large cities. To fit everybody in they built *insulae:* the ancient equivalent of apartment buildings, up to an unprecedented 10 storeys tall. (The pyramids were of course much taller, but you certainly couldn't live in them.)

Spreading across an entire block of the city, the *insulae* were surrounded on all sides by roads (appropriately enough, *insula* means 'island'). Instead of a central atrium for light and air, which was typical in most private homes at the time, the *insulae* had windows facing outwards at the city: in effect they were turned inside out. The first storey was built by installing many columns and then spanning relatively shallow arches between them. Concrete was laid over the curved tops of the arches to level them off and create a floor. Without the arch, far more columns would have been needed to support the floor beams, which would have created even tinier, more obstructed rooms.

To go higher, the Romans layered on more columns and arches. For the first time, they had to consider the design of foundations to ensure that their large, heavy structures didn't sink into the ground. After studying the type of earth present under the proposed building, they constructed foundations made from stone and concrete to hold the structure up.

The most expensive, sought-after apartments were on the ground floor. The higher you went, the smaller and cheaper the dwellings became – which is of course the opposite of today: the height of luxury (literally) is a penthouse that will cost you a small fortune. The *insulae* were rather harried places: without elevators, residents had to trudge the stairs to the upper floors. Since water couldn't be pumped that high, they had to lug clean water up with them, and drag their waste back down (although many would simply throw it out of the window). En route you might even encounter an animal: a cow is said to have wandered up to the third storey of such a block.

The *insulae* were noisy: even after glass windows were invented and replaced shutters, they couldn't keep out the constant commotion of Roman streetlife. Before dawn, the bakers were out clanging their ovens. Later in the morning, teachers would be shouting out their lessons in the squares. All day you could hear the constant hammering of the gold beaters, the jangling coins of the money changers, the cries of beggars and of vociferous shopkeepers trying to strike a bargain. At night, dancing, drunken sailors and creaking carts added to the din. But worse than the noise and lack of sanitation was the fear that your building might collapse or burn down, as happened to a number of poor-quality blocks. The emperor Augustus instituted an early form of planning restriction, limiting the maximum height to about 20m (later adjusted by Nero to just under 18m), but these regulations were often disregarded. Despite the discomforts, by AD 300 the majority of Rome's population lived in *insulae*. There were over 45,000 such buildings, and in contrast, fewer than 2,000 single-family homes.

For the first time in history, practical tall structures for hundreds of people, spread over many storeys, were built. It was a revolutionary idea – although it must have been a disconcerting experience for the first inhabitants, rubbing shoulders with their neighbours, and a bizarre sight for outsiders unaccustomed to this new way of living. This, though, was the future.

This idea – humans living in layers on top of one another – was the start of what would eventually become the skyscraper.

*

Archimedes took the Mesopotamians' pulley and improved it. Similarly, the Romans took Archimedes' innovation and applied it in new ways, creating heavy-duty cranes in the process. But advances in engineering don't come just from picking up a tradition or innovation and taking it forward. Sometimes they are about breaking with tradition and thinking the impossible. I admire Leonardo da Vinci (1452–1519), for example, who envisioned flying machines, mechanical knights and even a famous concept for a bridge (made from short ladder-like units that could be assembled and disassembled quickly). Another such thinker was Filippo Brunelleschi (1377–1446), who singlehandedly – and, as you'll see, single-mindedly – created one of the most famous domes in Renaissance architecture, and revolutionised construction in the process by building it without a supporting framework. Not bad for a man after whom people shouted, 'There goes the madman!'

By Brunelleschi's time, work on the Cattedrale di Santa Maria del Fiore in Florence had already been under way for more than 100 years. An edict of 1296 had proposed the construction of an edifice 'so magnificent in its height and beauty that it will surpass anything of its kind built by the Greeks and the Romans', and building began that same year, following designs by Arnolfo di Cambio (who was also responsible for two other great Florentine landmarks, the Basilica di Santa Croce and the Palazzo Vecchio). Despite the edict's grandiose assertions, enthusiasm and civic energy – not to mention cash

– waxed and waned in the following decades, and as a result it wasn't until 1418 that the cathedral was finished – except for its dome. During construction, little thought had been given to how someone might place a dome on what was, for the times, a massive hole of 42m.

Brunelleschi's Duomo in Florence, which caps the Santa Maria del Fiore cathedral in this Italian city.

Brunelleschi grew up close to the building site and its unfinished cathedral. Construction had been going on for so long that one of the streets by the site was now called Lungo di Fondamenti: 'Along the Foundations'. As an apprentice, he learnt to cast bronze and gold, forge iron and shape and form metals. He later moved to Rome to study the techniques of his ancestors, the ancient Romans. Brunelleschi had always been drawn to engineering and made two resolutions as a young man: to revive architecture to the greatness of ancient Roman times, and to provide a dome for the cathedral. The chance to fulfil both resolutions presented itself when the authorities in charge of the structure ran a competition to find a suitable candidate to build the dome. But Brunelleschi was unlikely to win unless he could overcome the hostility his radical ideas engendered in lesser imaginations, and diplomacy was not his strong point. (On one occasion a committee reviewing his designs had him forcibly ejected from their presence and thrown into the piazza, which is what earned him a reputation as a madman.)

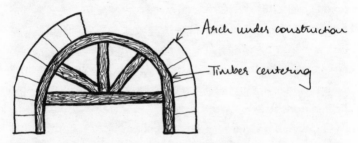

The construction process of building an arch whereby a timber centering allows the stones to be placed in position, finishing with the all-important keystone.

It's perhaps easy to understand why people denounced Brunelleschi's claims that he had a new method of construction. For thousands of years, arches – and domes – had been built in the same way. Carpenters made a timber template or *centering* to match the shape of the underside of the arch. Stonemasons or bricklayers carefully added material around this template, often gluing the masonry together with some form of mortar. They started by laying brick or stone from the base, working their way slowly towards the centre of the arch. The final stage was crowning the arch with a *keystone*. Until the keystone was placed, the curved arms that sprung up from the base remained disconnected. The timber centering supported them; without it the arch would simply have collapsed. Once the keystone was placed, the pathway for the compression loads was complete, and the arch was stable. The centering could then be removed and the arch would remain standing. The construction of domes followed the same process, but used a hemispherical timber centering.

Everyone believed this was the only way to build a dome. Brunelleschi disagreed. He presented a model to the committee that was 2m wide and almost 4m high, made from 5,000 bricks, which he said had taken just over a month to complete and had been built *without* using centering. The claim was met with scepticism, especially since he refused to tell anyone how he had done it.

The panel of judges tasked with choosing the dome's final design repeatedly asked him to reveal his methods, but Brunelleschi refused. At one of the judging meetings, where a number of experts were present and also bidding for the commission, he asked for an egg to be brought into the room. If any of his rivals could make the egg stand on its end, he said, they should win the competition. One by one people took the challenge, and failed. Brunelleschi then tapped the egg hard on the table and left it standing where it was (with a partially broken shell). When the others protested that anyone could have done that, had they known they could break the shell, he countered: 'Yes, and you'd say the same thing if I told you how I intend to build the dome.' He won the contract – though possibly only because there were few other practical solutions. (One person had even suggested filling the cathedral with earth to support the dome during construction. After the dome was completed, the earth would be cleared by small boys eager to get hold of coins deliberately mixed in at the outset.)

I visited Florence when I was a physics student. With the Ponte Vecchio, Giotto's Campanile, the Baptistery and Santa Felicita, it's like an open-air museum of medieval and early Renaissance engineering. *Il Duomo*, as the city's cathedral is affectionately known, is of course one of its centrepieces. I stood outside for a while, taking it all in – the neat symmetry of its three doorways, separated by four tall columns (with another two up above), and a series of very intricate carvings of Mary and the apostles just below the largest of the rose windows. Circles, pointed arches, triangles and rectangles, with coloured bands of stone, came together in pleasing geometric chaos. Eventually I passed through the doorway and my eyes were immediately drawn to the underside of the dome, high above me.

The base was an octagon, and each side had a circular stained-glass window letting in shafts of light. More light entered through an *oculus* at the top of the dome. Above the stained-glass windows were spectacular frescoes depicting The Last Judgement – choirs of angels, saints and personifications of the virtues vied for attention amid layers of painted cloud. It was all lovely, but the scientist in me wanted to know how it worked, to see the dome behind its beautiful embellishments.

The best view of the dome is from Giotto's bell tower, which stands in the piazza near the western corner of the cathedral. The 414 stone steps tested my fitness, but eventually I found myself at the top, looking out at the bank of deep red terracotta tiles and a few of the eight white ribs that define the dome's shape. It was a thrilling viewpoint, and a fitting tribute to Brunelleschi's genius. For me, it's Brunelleschi's unconventional thinking, coupled with the courage to make it a reality, that makes him relevant to modern engineering. It's by thinking beyond the orthodoxy and imagining the 'impossible' that we move engineering forward.

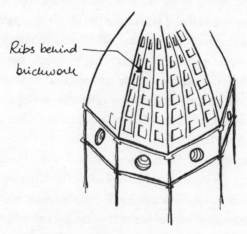

Ribs behind brickwork

The skeleton of the Duomo that lies between the two layers of brickwork, Brunelleschi's innovation.

Brunelleschi drew the ribs in characteristically detailed sketches. The ribs were made from stone, acting as arches that landed on the eight corners of the hole. These arches supported the edges of the octagonal dome. Between the main eight stone ribs were a further sixteen designed to resist the force of the wind. I couldn't see these from the outside, because Brunelleschi hid them away in the hollow space between two layers of brick skin. By creating this hollow space, not only was he able to hide the secondary ribs, he could also reduce the weight of the dome to half of what it would have been if it was solid. This reduction in weight helped him build the dome without centering.

Brunelleschi had gone back to basics. Brick structures are traditionally built in layers, comprising brick, then a layer of mortar, then another layer of brick, and so on. Imagine a simple garden wall and you've got the idea. Say, however, that you need this wall to curve in towards you (unlikely, I know, but bear with me). At that point, the problems begin: as the wall curves and becomes taller and heavier, it's in danger of overloading and cracking. Mortar is usually weaker than brick, so the continuous layer of mortar, rather than the bricks, is most likely to fail first.

To counter this, Brunelleschi asked his bricklayers to do something they had never done before. He directed them to lay three bricks horizontally, and then to place bricks vertically, like bookends, at either side of the horizontal group. The next layer again alternated three horizontal bricks with vertical bricks at each end. It was a painstaking process: four million bricks were laid; workers patiently waited for the mortar to dry on one layer before they started on the next. The layers created a 'herringbone' pattern, so-called because it supposedly looks like the bones of a fish. As an engineer, I admire this idea because of its simplicity. Since continuous lines of mortar were the weak link, Brunelleschi broke up the lines with vertical bricks, making the curving wall far stronger.

A herringbone brick-laying formation in which the vertically laid bricks add strength.

A similarly innovative approach drove the construction of The Shard. While designing its spine (or core), the team of engineers I worked with devised a unique method to build it. To save time on the construction programme, we decided to work in two directions: digging down to form the basement and at the same time constructing upwards. Usually when you want to make a basement, you dig an immense hole with concrete or steel walls holding up its sides. Piles – long shafts of concrete – are installed at the bottom of the hole to support the future building. Then slabs are poured at each basement storey until you get back up to ground level. It's only at this point that anything can be built above ground.

But we did something unprecedented. We asked for the piles to be installed at *ground* level, and huge steel columns to be plunged into the piles. First, the ground floor slab was built, with a giant hole in it. This hole gave workers access to the soil, then diggers removed earth to expose the concrete piles with steel columns inside them. While digging continued downward, a special rig was attached to the newly exposed steel plunge columns, this rig could build the central concrete core.

As the core rose, the basement and foundations were finished. At one point, twenty floors of the huge concrete spine were being held up just by the steel columns – there was no foundation in place. It was a structure on stilts.

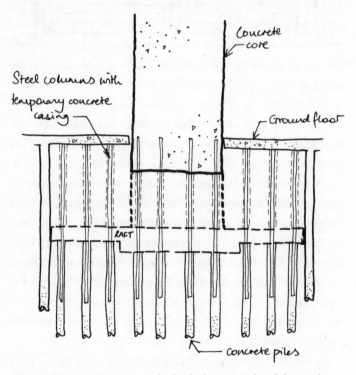

Concrete core

Steel columns with temporary concrete casing

Ground floor

RAFT

concrete piles

The top-down construction method, which was employed during the building of The Shard, London.

This method, called 'top-down' construction, had been used previously to hold up columns and floors in small structures. But it had never been used on a core, let alone one of this size. It was an engineering first. Our ability to think beyond standard practice saved time and money – we solved a real-world challenge with creativity. Others are now using our idea in their projects – as always, building on existing ideas leads to innovation, whether it's in one of the most famous cathedral domes in the world, or one of the tallest buildings in Europe.

*

On that site visit to The Shard in May 2012, as I shot up the tower in my cage-like hoist to the 34th and then the 69th floors, my eyes glued to the building rather than looking out and down, I couldn't help reflecting on how, without elevators, The Shard – indeed, any skyscraper – simply wouldn't exist. Part of the reason Roman *insulae* stopped at ten storeys was because climbing up and down any further was impractical. Today, we're so used to pressing a button and summoning a mobile cubicle to whisk us up and down our multi-storey towers that we don't give it a second thought. But before the 1850s, elevators in this form didn't exist. And although we started to build skyscrapers fairly soon after the invention of the elevator, such a device wasn't originally designed with buildings in mind, but as a safer way to move materials around a factory.

Like Archimedes, Elisha Otis had a restless and creative imagination. While working in a variety of jobs – carpenter, mechanic, bedstead manufacturer, factory owner – he invented an automatic turner that made the production of bedsteads four times faster; a new type of railway safety brake; and even an automatic bread-baking oven. In 1852 he was hired to clear a factory in Yonkers, New York and, frustrated by the effort involved in transporting materials between floors manually, he turned his attention to how best to accomplish the job mechanically. Methods for moving people and materials from one storey to another had been around for centuries: Roman gladiators, for example, rose from the pits of the Colosseum up into the fighting arena on a moving platform. The problem, however, was that they weren't safe: if the rope shifting the platform up or down suddenly snapped, the platform fell to the ground, probably killing its occupants. Otis wondered if he could fashion something that would prevent this from happening.

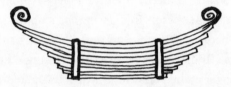

The wagon spring solved the challenges of operating an elevator.

His idea was to make use of the 'wagon spring': a C-shaped spring made up of carefully layered thin steel strips that was commonly used to improve the suspension in carriages and wagons. When it has force on it, a wagon spring is almost flat, but when it's released, it curves. It was this change of shape, caused by force, that Otis planned to use to his advantage. First, he replaced the smooth guide rails (which kept the platform in position during its progress up and down) with toothed or ratcheted rails. Then he created a mechanism in the shape of a goalpost, which had a hinge in the middle and feet sticking out at the base. He attached the spring, then the goalpost, to the rope at the top of the elevator car. When the rope was intact, the spring remained flat and the goalpost square. If the rope was cut, the spring sprung into a C-shape, pushing down on the goalpost and deforming it so that its two 'feet' stuck into the ratcheted rails, bringing the elevator to a halt.

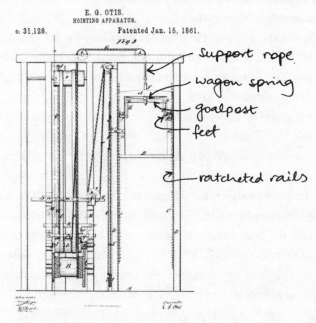

E. G. OTIS.
HOISTING APPARATUS.
o: 31,128.   Patented Jan. 15, 1861.

support rope

wagon spring

goalpost

feet

ratcheted rails

This diagram is included in the patent documents for the Otis Elevator – or 'hoisting apparatus'.

But to bring his invention to the attention of the public, and show them that it worked, Otis needed a big stage – and he found it at the 1853 World's Fair in New York. Entitled the 'Exhibition of the Industry of All Nations', the exposition aimed to show off American technological might, and showcase industrial innovation from around the world. In the vast exhibition hall Otis constructed his elevator with guide rails, ratchets, springs, platform and hoisting machinery, and loaded the platform with goods. When a crowd had gathered, he climbed on top of the platform and had it lifted to its maximum height. As the crowd looked on, he called for the hoisting-rope to be cut, and his assistant swung the axe.

There were gasps as the platform suddenly lurched downwards. And then, just as suddenly, it stopped. It had fallen only a few inches. From the top of it Otis could be heard shouting, 'All safe, gentlemen. All safe.'

Four years later, Otis installed his first, steam-powered safety elevator in the five-storey E.V. Haughwout & Co. department store on the corner of Broadway and Broome Street in New York. The eponymous company he founded has continued to supply elevators and escalators to buildings around the world, from the Eiffel Tower and Empire State Building to the Petronas Towers in Malaysia. Such buildings would hardly have been possible without Otis's invention. Until he developed the safety elevator, the height of a building was restricted by how many stairs people were prepared to climb. The elevator smashed that barrier and engineers could start to think about true skyscrapers.

Since then we've been building higher and higher, and we now have the opposite problem: we can't make elevators that travel much further than 500m because the steel cables to hoist them up and down become too heavy for the machinery to work efficiently. It's one reason why elevators often don't go all the way to the top of very tall towers. You go up a number of floors, then change elevators to go up the rest. But engineers are already exploring ways to solve this by using different materials. Replacing steel with carbon fibre – which is stronger but lighter – seems one way forward, but questions remain about how well the carbon fibres can resist fire. As our towers continue to grow, these innovations will be much needed.

Another challenge in super-tall towers is sway. In chapter 1 I talked about controlling the movement of buildings to stop us feeling sick. But there is another reason this control is needed. Elevators run on straight guide rails, and as towers move the elevator shafts and the guide rails fixed to them curve. A small amount of curve is not a problem – the cogs and clasps of the elevator car on the rails have a little give – but too much and the car will grind to a halt, unable to move. The taller buildings become, the more they move and the more curve you experience in the elevator shaft. There are solutions to the problem, ranging from upgrading the elevators themselves, to allowing more give, to stopping elevators running in the worst storms. Ultimately, I'm sure, a modern-day Otis will come up with an ingenious solution. And he – or she – will have to, because the elevator has become an intrinsic part of our everyday life. The equivalent of the entire world's population is moved in an elevator every 72 hours.

*

I was reminded of Elisha Otis during my visit to the Burj Khalifa in Dubai, the world's tallest building (at 829.8m), because his company installed the elevators that were about to take me to the observation deck on the 124th of its 163 floors. It was a more serene journey than my trip up the outside of the tallest tower in Western Europe in a cage-like hoist, although the floor number on the LCD display changed with a bewildering rapidity as we ascended at 36km/h.

(Elisha Otis's original elevator in the E.V. Haughwout Building climbed at just over 0.7km/h.) A minute later I emerged to an unparalleled view. On one side, pure sand extended beyond the buildings to the horizon. On the other, I could see the blue sea and, far away to the left, the cluster of man-made islands that form the famous leaf shape of the Palm Jumeirah. Steeling myself, and feeling protected by

Burj Khalifa in Dubai, the world's tallest building in 2018, which has been made possible partly by the developments in elevator technology.

the floor-to-ceiling glass, I ventured closer to the edge and looked down. Beneath me were a number of tiny, futuristic-looking buildings, like scale models on the set of a sci-fi film. It was a shock to realise that these structures are actually taller than most of the skyscrapers in Europe, and many even in the US. The Burj Khalifa dwarfs everything around it, and plays havoc with your sense of proportion.

'Megatall' skyscrapers like the Burj Khalifa were made possible by a man who started life as a mischievous and lively-minded young boy, born in Dhaka, Bangladesh, in April 1929. Fazlur Khan disliked traditional schooling methods: his inquisitive questions were met with stern responses from teachers; as a result, he didn't take education very seriously (even though his father was a mathematics teacher). Fortunately, his patient, forward-thinking dad realised that his son needed a broader education, and was determined to further his intellectual curiosity while fostering a sense of discipline. He set Fazlur problems similar to those in his school homework, but which made the boy consider solutions far beyond what the homework asked for; he also challenged him to solve the same problem from multiple perspectives. When the time came for Fazlur to choose whether to study physics or engineering at university, his father guided him towards the latter because, he said, it demanded discipline and would require him to wake early for lectures. (In fact, as I can attest, a physics degree involves a lot of early-morning lectures too.) Khan gained a degree in civil engineering at Dhaka University in 1951, finishing first in his class, and went to the US on a Fulbright Scholarship in 1952. In the next three years he acquired two master's degrees and a PhD, while also learning French and German.

It was Khan who came up with the idea of putting a building's stability system on the outside – a brilliant innovation that has since been used on iconic structures around the world, from the Centre Pompidou and the Gherkin to the Hearst and Tornado Towers. Using large pieces of diagonal bracing to form strong triangles, Khan created a stiff external skeleton, effectively turning traditional skyscrapers inside out. This system is often called a 'tubular system' because, like a hollow tube, the outside 'skin' of the structure gives it strength, although the shape of the skin doesn't have to be cylindrical.

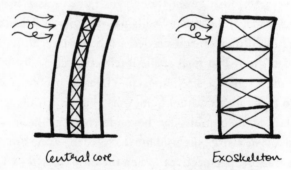

Central core   Exoskeleton

An alternative stability system for buildings is to forgo the conventional central core and instead employ an exoskeleton.

Khan's first commission to employ this concept was the DeWitt-Chestnut apartment building in Chicago. But the real showcase for his novel approach was the completion in 1968 of the city's John Hancock Center which, at 100 storeys (344m), became the second-tallest skyscraper in the world after the

Empire State Building. It is a rectangular cuboid with gently tapering faces, making it narrower at the top than at the base. On each face you can see five giant 'Xs', one on top of the other, that form the bracing for the tower. Fifty years on, its eye-catching design still looks modern and elegant. The pioneering design earned Khan the catchy title 'father of tubular designs for skyscrapers'.

The John Hancock Center in Chicago utilises an exoskeleton to give the tower stability.

The external skeleton was only one of Khan's ideas. He also suggested combining many such skeletons in a cluster. This is like holding a bunch of straws in your hand: each straw is a single tube which by itself is stable up to a certain point; by bunching lots of straws together, however, you can make a much stiffer and more stable structure. The Burj Khalifa employs a variation of this system. Look at a cross-section of the structure and you'll see that it has a distinctive tripartite shape that resembles leaves or petals. (It's become a kind of brand image for the building: as you ascend in the elevator, a lightshow of row upon row of the shapes dances across the walls in different configurations.) The 'petals' are in fact a series of 'straws' or tubes with exoskeletons of their own which – in their cluster – support one another. This mutual support between the individual pieces means that the tower remains stable despite being so high.

The key to building higher is stabilising the structure externally rather than internally. Perhaps the most precarious experience I can think of was the only skiing trip I ever went on. At first, our instructor wouldn't let us use ski poles, so I had to stop myself from falling over using just my feet. I soon lost count of the number of times I fell over, and the number of bruises I picked up in the process, but once I had managed to stand upright – at least for a short while – I was allowed the poles. And what a huge difference they made: by spreading my arms out and using the poles to stabilise myself, I found I could stay upright longer. Although the poles were much skinnier and less stiff than my legs, by putting them further apart than my feet could reach, I was more stable.

Tall towers with exoskeletons work in the same way: by spreading the stability from a small internal area (analogous to my feet or a building's core) to an outside area (the poles or the exoskeleton), it's possible to create much more stable buildings. Flipping the structure around in this way opened up a number of engineering possibilities: if you built a tower of 50 or 60 storeys like engineers did at the turn of the twentieth century, you could use much less material, making it cheaper. Or if you used the same amount of material as in the older towers, you could build much taller. So, from the 1970s onwards, scores of tubular towers arose, from Hong Kong's Bank of China Tower and the original World Trade Center Towers in New York to the Petronas Towers in Kuala Lumpur, changing the face of our skylines forever and creating the classic modern-city silhouette.

*

With the invention of new building techniques, structural systems, and computing power increasing every year, it's an exciting time to be a structural engineer. Just as the height of buildings has increased on the back of what we've learnt from our predecessors, so too has the depth of our knowledge. Today, I can design structures that brilliant thinkers like Leonardo da Vinci struggled with. And in a hundred years, engineers will no doubt find it easy to do things that I struggle with now. My peers and I are building on thousands of years of engineering gifted to us by Archimedes, Brunelleschi, Otis, Khan and countless others.

With today's technology at our fingertips, I don't believe there is a limit to how high we can build. We've beaten so many physical, scientific and technological restrictions over the past 4,000 years that with strong enough materials, a wide enough base, solid enough ground – and, I suppose, enough money – I see no reason why we can't go as high as we want. The real question is: how high do we want to go? A wide base would probably mean very little daylight in the middle of the vast floors. Strong large columns and beams could mean restricted spaces in which to live and work. And what about the safety and convenience of the inhabitants: how long would you need to wait for an elevator, and how would you evacuate tens of thousands of people from a mammoth building?

Technology can undoubtedly take us there. New super-strong materials like graphene are already being synthesised in labs; cranes are getting larger; and new techniques like top-down construction are constantly being used in inventive ways. Science and engineering are leading to the creation of the mega-skyscraper – the Wuhan Greenland Centre (636m) in Wuhan, China; the Merdeka Tower (682m) in Kuala Lumpur, Malaysia; and the dart-like Jeddah Tower in Saudi Arabia, which will be the world's first building to reach a height of 1km – at an unprecedented pace.

But where does it all stop?

The highest I've lived is on the 10th storey, and I loved the view and the new perspective of the city in which I lived. But I wonder how I would feel living much higher than that. In cities like Hong Kong or Shanghai, living on the 40th floor is common for thousands of people: it's something the residents are used to. Eventually, perhaps, it will be commonplace every-where: people are moving to cities in droves, and building high is a good way to fit all of us into an increasingly limited space.

The rapid growth in the height of buildings in the last century has barely given us a moment to consider if we *like* being so high above ground. But now, rather than racing ever higher, we are now stopping to think about our desires. It's about what we *want* to build, not what we *can*. After a spate of building high towers from the 1960s to the 1980s, architects and engineers are questioning what type of buildings are really best for people and the environment. Cultural factors also play a part: different countries are at different stages in their urban development, and can have very different views about whether onwards and upwards is the best approach. I believe that, at some point in the future, the average height of our towers will plateau. Sure, iconic towers will still be built and they will continue to break records. Ultimately, however, our humanity will hold us back from the mega-tall. We want to live with sunlight and air flowing into our homes, and a connection to the earth and to our roots. We might gaze upwards at our structures and marvel at them, but we also need to feel grounded.

EARTH

Mexico City is built on a lake.

It started off as a small island but gradually expanded. The city now spreads far beyond its original site, but the centre of town, which contains most of the historical Aztec and Spanish buildings, sits on that lake. Twenty-eight metres down, the earth is strong and solid; everything on top of that is loose soil that was added later, and the result is very soft, very wet and very weak. It was described to me as a 'bowl of jelly with buildings on top'.

Mexico City, which is built over a lake.

And so the historical centre of Mexico City is sinking. Fast. In the past 150 years it has subsided by over 10m – that's more than a three-storey building.

*

When I was invited to Mexico to give a talk about my career and designing tall buildings, I jumped at the chance, not least because there was so much I wanted to see: the National Museum of Anthropology, the Bosque de Chapultepec, the ancient pyramids at Teotihuacan, and of course the Torre Latinoamericana, once the tallest skyscraper in Mexico City and still one of the best places to appreciate the sheer sprawling vastness of the metropolis. Naturally, I was also keen to explore the unique ground that lies below the city, and the bizarre effect it has had on the buildings there.

In engineering, what lies beneath the surface is just as important as what we can see above it. After all, you can have a well-designed superstructure (the bit above ground) – but if it's not supported by an equally well-designed, stable substructure (the bit below the ground); if the layers and condition of the soil being built on aren't properly understood; if you don't build correctly within that ground – then the structure won't be stable. The end result could be the Leaning Tower of Pisa. (Not the reason I would want tourists flocking to one of my buildings.) Knowing that Mexico City has some of the most challenging ground conditions in the world for building on – plus seismic susceptibility for good measure – I figured my trip was a fantastic opportunity to hear directly from the experts how they keep the city standing straight.

The site of the city was determined by a vision. The Aztecs were told by their god Huitzilopochtli (the God of War and the Sun) that they must move from their highland plateau, and that their new capital must be located where they found an eagle with a snake in its beak sitting on top of a nopal cactus (an image that is now the emblem on the national flag). The Aztecs set off and, after searching for just over 250 years, they found the eagle their deity had foretold. The fact that it was sitting on a tiny island in the middle of Lake Texcoco didn't seem to trouble them (although I can imagine the tribe's engineers cursing under their breath as they surveyed their new, watery building site).

Tenochtitlan, which means 'place of the nopal cactus', was founded in 1325. In its heyday, it was a beautiful city with fertile gardens, canals and massive temples, and its rulers commanded vast swathes of land. To connect the island city to the mainland, the Aztecs built three large causeways by pushing wooden logs vertically into the lake, and then creating pathways on top with soil and clay. These causeways are now the main roads that run through the historical centre of the modern city.

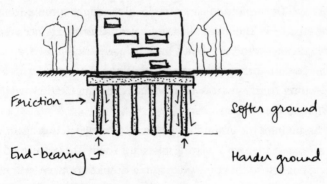

Piles holding up buildings in soft ground.

The logs are examples of *piles*. They come in various shapes and sizes but share a common principle: they are columns put deep into the ground to help support the structure above them. If the ground is soft and not strong enough to support the weight of the structure, piles work to channel that weight in such a way that the soil is not overwhelmed. The ancients generally used tree trunks, but modern piles supporting larger structures are usually made from concrete shaped into cylinders, and sometimes from steel, cast in circular tubes, H or trapezoidal shapes. The foundations of the structure are built at the top of these piles and connected to them through steel bars.

Piles can channel forces into the ground in two ways: by means of friction between the surface of the piles and the soil, or by dumping forces at their base ('end-bearing piles'). Depending on the weight and type of structure being supported, you can have multiple piles, which can vary in length depending on the forces they feel and the type of ground they engage.

Friction piles exploit the friction between the surface of a pile and the ground to carry the load or weight coming from the structure. The more piles you have, the more surface area is in contact with the ground, and the more friction is created. This friction force resists weight – thinking about it in terms of Newton's Third Law, it is an upward reaction to the downward action of the superstructure.

Sometimes the ground is too loose to create friction against a pile, and then end-bearing piles are used. These are made long enough so that they poke into a deeper, stronger layer of ground. The load in the piles flows into their bases and dissipates into the earth.

In fact, piles don't have to be either friction piles or end-bearing ones: they can be both. Some soils, such as clay, have good friction capacity because they bond to the pile. But say the load is so large, and you're so restricted by available space, that friction alone isn't enough to resist it. In that case you can make the piles long enough to reach a stronger layer of ground. In London, for example, there is a highly compact layer of sand approximately 50 m deep that we drill down to for larger structures.

Working out how many piles to use, and how big to make them, is an important part of the engineer's job. The starting point is the soil-investigation report, which tells me what the different layers of ground are, and how thick and strong they are. Then, if I find that a 'pad' of concrete will not be enough to stop the structure sinking, I'll choose to use piles. By consulting the information in the report – and geotechnical engineers – I can calculate how deep the pile needs to be to hit a strong layer, and what the friction properties of the various layers are.

I then have to decide on diameter. A small-diameter pile has the benefit of being cheaper and easier to install, but it may not be strong enough for the job. A larger-diameter pile has a bigger surface area, which increases the amount of friction; the area of the base is also bigger, making it stronger. The calculation is a search for the right compromise. I choose a diameter, calculate how much load a single pile will take based on a chosen length, then divide the total weight of the building by the capacity of one pile to work out how many piles I need. If I can fit that number of piles below the structure, then we can go ahead. If not, I make the pile bigger and repeat the calculation. For a 40-storey tower I designed near Old Street in London, we arrived at a total of about 40 piles between 0.6m and 0.9m in diameter, with some more than 50m long where the loads were greatest. Many modern skyscrapers are held up by piles that work by friction alone (if the ground is good enough so the piles can carry the loads they need to). But the piles in this tower work both by friction and by using end-bearing, as London's clay is relatively weak to quite a depth.

Putting piles in the ground is a big challenge in itself. It wasn't really until modern mechanisation that the huge piles we can now install were possible. Now, piles are often built using a sort of giant corkscrew that twists deep into the ground then reverses out, bringing the soil with it, and leaving a hole that is later filled with concrete. While the concrete is still wet, a steel cage is plunged in to reinforce the pile. For centuries, before mechanisation, most engineers simply pushed piles into the ground, as the Aztecs did at Lake Texcoco. From an engineering point of view their construction was successful, standing firm for the next two centuries.

But then the foreigners arrived.

The Spanish captured Tenochtitlan in 1521, razed it to the ground, and then rebuilt the city on the foundations of the Aztec pyramid temples. They cut down trees around the lake, causing mud slides and erosion that made the lake bed shallower. The water levels rose and the city flooded frequently throughout the seventeenth and eighteenth centuries, causing chaos and devastation (after the flood of 1629 the city was underwater for five years). Eventually, the lake was filled with soil to allow the city to expand, but it still suffered regular flooding because of the high level of water naturally present in the ground.

There is a level in the ground below which natural water flows and saturates the earth: this is known as the *water table*. Dig a hole in an area where the water table is high, and you'll find that the hole fills with water pretty quickly: this is like the original Lake Texcoco. If you fill the hole with earth – which is like Lake Texcoco being filled with soil – then sprinkle on water to simulate rain, eventually water will puddle above the soil (just as our gardens are covered with puddles after a storm because the soil is saturated). This is what happened in Mexico City. The lake was filled in with soil but the water had nowhere to go. Then, the moment it rained, the rain added to the underlying water table and stagnated in the streets of Mexico City. It wasn't until the twentieth century that the flooding was controlled using a huge network of tunnels that led the extra water away. But the legacy of building on such unpredictable, unstable ground can still be seen in the modern city.

*

Standing in the courtyard outside Mexico City's enormous and very grey Metropolitan Cathedral, I scanned the crowds for Dr Efraín Ovando-Shelley, a geotechnical engineer who, according to his photo, wore sunglasses and khakis that made him look a bit like Indiana Jones. The solid, ordered columns of the cathedral were in sharp contrast to the delicate carvings between them, but what really caught my engineer's eye were the cracks in the building. I could see where black space had opened up in the mortar and stone bricks, and the two huge bell towers that flanked the main entrance didn't seem to be completely vertical. But such considerations were cut short when, at exactly the appointed time, Dr Ovando-Shelley appeared wearing his sunglasses, greeted me, handed over a book he had written, and led me towards the cathedral for a very unusual guided tour.

Metropolitan Cathedral, Mexico City.

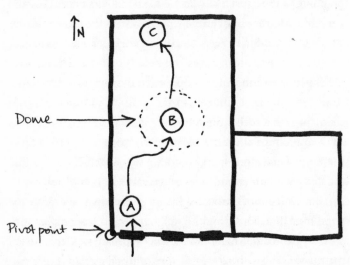

Map of the Metropolitan Cathedral.

As soon as we stepped through the entrance, *(Map, point A)* something felt very odd to me. Swarms of tourists stood rapt by the grandeur of the place, while worshippers sat respectfully hunched in its polished wood pews. But my attention was drawn to the floor. As we moved towards the back of the cathedral, I felt like I was walking uphill. And I was – because of the uneven or 'differential' settling of the ground that has taken place through history, the floor of the cathedral slopes upwards.

Construction of the cathedral began in 1573, on top of the foundations of an Aztec pyramid. The architect, Claudio de Arciniega, knew of the problems with the ground and designed a clever foundation to deal with them. He started by driving more than 22,000 wooden stakes – each 3m to 4m long – into the ground, to 'pin' the soil together and compact it. Imagine a box of sand with lots of kebab skewers pushed into it in a grid pattern. If you shake the box, you'll find that the sand moves around far less than if the skewers aren't there. The stakes performed a slightly different function from piles, since they weren't designed to take the weight of the cathedral, but rather to strengthen the soil.

Following this, the builders erected a massive masonry platform above the stakes. It measured 140m by 70m – about the same width as a soccer pitch but one and a half times longer – and was about 900mm thick. Huge beams were laid on top of this platform in a grid pattern – a bit like a waffle – in such a way that the columns and walls of the cathedral could sit on top of them. The tops of the beams would eventually form the floor of the cathedral, spreading the weight of the columns onto the masonry platform, which in turn would spread the weight over the ground. This sort of foundation (with or without the large beams) is known as a 'raft' foundation.

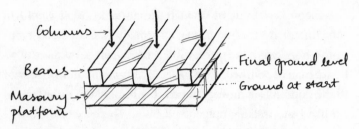

The layers that form the raft foundation of the cathedral.

It does what its name suggests, which is to 'float' on top of the ground. When building on soft ground, the key is not to put large concentrated loads on the soil. If you do, it's like standing on mud in stiletto heels. As many summer wedding guests will know, a sharp heel sinks into the ground because the pressure it exerts on the ground (calculated by dividing force by area) is high. Flat shoes, however, don't sink as easily because the same force is spread over a much larger area – the snowshoe is based on this principle. So the masonry platform in the cathedral acted like a flat shoe on top of mud, spreading the weight of the building over a large area. The trouble, however, is that sometimes the ground is so soft that even spreading the weight of a structure across a large area, and avoiding concentrated loads, is not enough.

It's probably worth noting here that friction or end-bearing piles were not used to support the weight of the structure. Perhaps because of the pyramid foundations below it, or perhaps because the engineers of the time realised that anchoring piles to the solid layer of earth might cause the opposite problem, making the cathedral rise. In fact, the Angel of Independence victory column in

Mexico City (built in 1910) is supported on piles, and in the 100 years that have passed since it was built, 14 steps have been added to its base as it has become taller relative to its surroundings. Engineers in Mexico City agree that it's best to allow the city's structures to slowly, steadily and uniformly sink.

When it was built, the top of the masonry platform was made level with the ground outside. On top of it were the 3.5m-deep beams, and on top of them was the floor of the cathedral itself. Thus the floor was originally constructed 3.5m above the ground, showing that the engineers knew the structure would sink, and planned that by the time they had finished it would sink just enough to bring the floor of the cathedral down to ground level. The hope was that the structure would sink uniformly, and wouldn't necessarily be damaged. Despite de Arciniega's efforts, during construction, as heavy stone was laid on top of heavy stone, the structure started to sink in a non-uniform way. The south-western corner of the structure (the front-left corner in the diagram) sank more than the north-eastern corner. To compensate for this unsettlingly uneven settling, the builders actually increased the thickness of the 900mm masonry platform on its southern side.

The structural reason why the platform settled unevenly is because soil comes with baggage. It's not enough to meet the soil, ask how it's feeling on the day you start building and then assume it doesn't have any emotions from its past that will affect how it behaves. It has a history and a character that an engineer must consider. The Aztecs had built their pyramid in exactly the place where the cathedral was sited, adding layers to it over time, partly for spiritual reasons, and partly to cover the damage caused by settlement. This construction had affected the physical state of the soil: some areas had already experienced lots of pressure and become consolidated and compacted, while others, which hadn't been weighed down, remained light and less dense. Where new foundations were built on top of consolidated soil they didn't sink much, but the portion built on less dense soil moved much more.

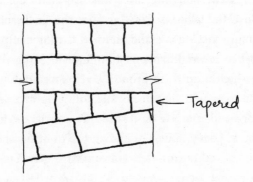

An attempt at realignment.

Even after the Spanish builders had finished the foundations, the structure continued to move unevenly. They tried to compensate for this differential settlement by changing angle as they worked up. Dr Ovando-Shelley pointed out areas where the courses of stone (which would normally be laid flat and uniform) had been cut to a taper. This helped the builders come back to a level line after the layers of stone they had already built had tilted. Other adjustments had been made to counter the continuing subsidence: a column at the southern end of the structure was almost a metre taller than the columns in the north. The cathedral was finished 240 years later, but throughout this time, and beyond, it continued to move erratically.

Dr Ovando-Shelley and I walked along one of the aisles *(see Map on p. 152, point B)* and stopped directly below the central dome. From here hangs a giant, missile-shaped pendulum (or *plumb line*) made of gleaming brass and steel that shows how far the cathedral has shifted. You can simulate this with a string, a small weight and a clear plastic box. Attach the weight to the string, suspend it from the centre of the roof of the box and lay the box on a level table top. You'll see that your makeshift pendulum hangs exactly above the centre of the floor of the box. If you tilt the box slightly, however, the pendulum will move away from the centre. Tilt the box by 45 degrees and the pendulum will hang over the edge of the floor. The Metropolitan Cathedral's pendulum works in the same way: as the foundations tilted, the pendulum stayed vertical. By noting where the pendulum was centred at various intervals over time, the tilt of the cathedral has been monitored.

In 1910 measurements were taken to compare the levels of the two extreme corners. The engineers established that, since 1573, the floor had tilted so much that one corner was a staggering 2.4m higher than the other. It's difficult to imagine a structure tilting by such an extreme amount; not surprisingly, it had a damaging effect on the cathedral's integrity. By the 1990s its bell towers were leaning precariously and in danger of collapsing.

A major restoration project started in 1993; Dr Ovando-Shelley was one of the large team of engineers that worked on it. They accepted that it was almost impossible to stop the structure sinking altogether, but reasoned that if it sank uniformly it would suffer less damage. However, before they could even think about ensuring it settled evenly, they needed to pivot the entire cathedral so it was relatively flat.

As my tour continued, we walked away from the dome to the back of the cathedral *(see Map on p. 152, point C)*. Here, the shimmering Baroque magnificence of the golden Altar of the Kings extended towards the ceiling, covered by a mass of intricate hand-carved figures – an opulent wall of worship designed to assault the senses, to impress, and to arouse reverence. It certainly inspired a feeling of awe.

I, however, was completely transfixed by a tiny metal stud on a column just to the left of the altar. It was relative to this point that the team measured and compared the levels of the floor to establish exactly how much the cathedral needed to be pivoted. The chosen pivot point (the point that wouldn't be allowed to sink any further) was the south-west corner, because this had sunk the most over time. The metal stud was at the northern end of the cathedral, which needed to be pushed down by metres. Just thinking about it made my head spin. And it didn't stop spinning as Dr Ovando-Shelley explained the technique they used to achieve it. Have you seen the sci-fi blockbuster *Armageddon*, in which Bruce Willis and his team must drill a hole in an asteroid and pack it with explosives to prevent a collision with Earth? The plan devised by the cathedral's engineers seemed about as unlikely and difficult to achieve: they would burrow beneath the cathedral and settle the soil. The thought of *removing* earth from underneath a structure to *stabilise* it might seem totally counter-intuitive. But for these exceptional ground conditions, exceptional engineering was needed.

As I said before, though, soil isn't just soil: you have to understand its history before you can predict how it will behave in the future. Dr Ovando-Shelley and the team performed a variety of soil tests all over the site to find out exactly how strong or weak the soil was, and how consolidated (or squashed down). Feeding this information into a computer model, they drew a 3D map composed of layers of different colours that undulated and overlapped depending on the strength and type of soil at a particular depth. The model also simulated all the historical events that had affected the soil – from the building of the Aztec temple and the Spanish cathedral to the changes in water level and so on – and created a profile of the ground.

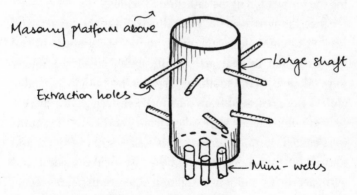

Extraction holes radiating from the large shaft.

The team then bored 32 cylindrical access shafts, 3.4m in diameter and between 14m and 25m deep, through the original masonry raft of the cathedral and into the ground. These were dug laboriously by hand (accessing this confined space with diggers would have been both difficult and dangerous). At each stage of the descent, concrete was cast in a ring around the edge of the hole, creating a tube to keep the soil in place. When the shaft was finished, a second layer of concrete was cast inside the tube to stop the hole from collapsing in on itself. At the base of each shaft the engineers sank four mini-wells from which they could pump out the excess groundwater that would otherwise rise and flood the shafts.

These, though, were not the shafts that were going to save the cathedral. They just provided the means for drilling about 1,500 holes, slightly inclined from the horizontal, with a diameter the size of a fist and between 6m and 22m long, through which soil could be extracted. The plan was that, after the soil had been removed, these holes would naturally close up over time, causing the foundation of the cathedral to settle.

Since the north side of the cathedral was the highest and needed to come down the most, the largest amount of soil was extracted in that area, while much smaller amounts were taken from the south-west corner. More than 300 cubic metres were removed from one north-eastern shaft, whereas only 11 cubic metres were taken from another in the south-west corner. In total, through this vast warren of shafts and tunnels burrowed deep beneath the historical cathedral, and with nearly 1.5 million extraction operations, 4,220 cubic metres of soil were removed from underneath the structure – enough to fill about one and a half Olympic-sized swimming pools.

As you might expect, this soil removal was done carefully and cautiously, in stages, over a long period (four and a half years). All that time the levels in the cathedral were strictly monitored to make sure that any movement stayed within the limits of what the engineers wanted. The arches and columns inside the cathedral were supported with steel beams and props to prevent any damage from sudden, unexpected or large movements. Meanwhile, soil samples were continually taken out of the ground to be tested for stiffness and water content, and were compared with the computer model to make sure reality matched prediction.

The difference in floor level between the north-east and the south-west had been more than 2m, but in 1998, once the north end had settled down by just over a metre, the process was suspended. Even though this left the foundation slightly tilted, the engineers had become concerned about damaging the structure. The lean of the towers had been brought back to an amount that was deemed safe – and so, for the time being, work has stopped.

The large cylindrical access shafts have been left open. They are now flooded with groundwater, but if they are needed in the future – if the cathedral starts tilting again – the water can be pumped out, and more soil removed. For now, the cathedral has been left to the mercy of the soil – but this time it is being watched.

Positioned at strategic points around the cathedral are four pendulums encased in glass boxes that send data wirelessly to a lab in Italy where engineers monitor how the structure is behaving. Pressure-pads monitor the loads in the columns, checking they aren't changing too much. A change in load would suggest the structure is tilting again, causing some columns to be more squashed than others. Dr Ovando-Shelley described the cathedral as a laboratory, in which data has been collected for nearly twenty years. It has become a place of science as well as a place of worship.

Since the 1990s, the cathedral has been sinking at a rate of about 60mm to 80mm per year – a slow and steady settling in comparison to the past and, most importantly, an almost uniform one. The movement will continue in the future, but it might slow down over time. This Indiana Jones of engineering had saved his relic, and succeeded in his mission. No Armageddon for Mexico City's Metropolitan Cathedral.

The team of engineers' groundbreaking work has been a subject of study all over the world. In 1999 they worked with engineers in Italy, replicating their methods below the Leaning Tower of Pisa. In Mexico City the engineers were faced with an extreme situation – the decidedly poor condition of the soil, its variability and the sheer size of the cathedral. But the upside of the challenge they faced is that we now have an invaluable body of knowledge that can be used by engineers in the future, particularly those fighting to save our heritage, and those attempting to build in harsher and harsher conditions as our population expands and the climate changes.

Our technical tour done, Dr Ovando-Shelley and I left the cathedral in search of a restaurant for lunch, crossing the Zocalo Square, which was framed by other elaborately designed and decorated buildings that had settled unevenly. He waited patiently as I stopped to take photos of door frames that had skewed from rectangles into parallelograms.

On a terrace overlooking the Zocalo, a waiter served us frozen margaritas. 'Soils have no word of honour,' said

Dr Ovando-Shelley, clinking my glass, 'and neither do geo-technical engineers.' He laughed uproariously. But to me, he had nothing but honour. He, and the team of engineers, had saved the biggest cathedral in the Americas from ruin. And he bought me chicken *mole* for lunch.

HOLLOW

Usually, our homes are an amalgamation of materials – we gather stuff and assemble it, creating something from nothing. But there is a place, with sparsely grassed steppes as far as the eye can see, where shelter was formed the other way round, in an absence of material – where nothing was created from something.

Naturally, I had been very curious to see this, which is why one day I found myself doubled over, surrounded by blackness, craning my neck and straining my eyes, trying to work out where I was. I knew I was deep underground: I had walked down hundreds of winding and incredibly steep stone stairs, past ancient living rooms, kitchens – and death traps – to get there.

I could just about make out that I was in a tiny, coffin-shaped passage, as wide as my shoulders as I crouched, and as wide as my feet at floor level. I wasn't even sure there was enough space for me to turn around and backtrack to the entrance. I could see damp beige stone just ahead of me, but the bright beam of light from my phone torch barely penetrated the darkness beyond. I carefully felt my way along the passage, trying not to bump my head. After what felt like a very long time (though it was probably only a few minutes), I emerged into a small lit cave and felt relief, until I saw the long rectangular recesses carved into the floor – which had once held the remains of those unlucky enough never to find their way out.

I was in Derinkuyu, one of the deepest and largest of the mysterious, warren-like ancient underground cities in the heart of Anatolia in modern-day Turkey. These cities were made possible by the area's three volcanoes – Erciyes, Hasan and Melendiz Daglari – which erupted violently around 30 million years ago. They spread a ten-metre layer of ash across the region, on top of which flowed lava, which consolidated and hardened the ash, turning it into what is known as *tuff*. The local climate, with its heavy rains, sharp changes in temperature, and melting snow in the spring, gradually eroded the soft tuff until only columns of it remained. The harder lava layer on top of the softer tuff degraded more slowly; now large pieces of lava rock sit precariously on top of the thin ash pillars, giving them a surreal, mushroom-like appearance – and

Fairy chimneys, the local name given to the thin ash pillars and the harder lava layers that sit precariously on top.

their local name: 'fairy chimneys'. The strange landscape acts as a kind of taster for the even stranger things going on below ground.

Geographically, Anatolia stands at the intersection of East and West, and throughout its turbulent history it has been the site of battles between civilisations. The Hittite people occupied the region in around 1600 BC, followed by the Romans, the Byzantines and the Ottomans. The constant warring meant that the locals were always under threat. The Hittites realised that the thick layer of compressed ash beneath their feet was relatively soft, soft enough to carve with a hammer and chisel. They began constructing underground caves and tunnels to hide in while the fighting went on above. Each of the civilisations that followed the Hittites added to these networks, in effect establishing cities in which up to 4,000 people could live for months at a time. Over a period of nearly 3,000 years, hundreds of underground cities were created in the region. Most of them were small, but about 36 had at least two or three storeys.

As I could see at Derinkuyu, the system of caves in these underground spaces was structured like an ant-house: the rooms were not stacked one on top of the other, as happens in our buildings, because that would cause the ash to weaken and collapse. Instead, the rooms were carved out randomly in space, spread out across a large area. The arched ceilings over the rooms and passages were the perfect shape to keep stone in compression, and stable, ensuring that the ground would not cave in on them. A number of ventilation shafts, starting from the surface and running for up to 80m underground, brought in fresh air. The cities were designed to protect against infiltration by the enemy – with huge rolling stone doors to keep them out, deep pits for them to fall into, and cubby-holes behind doors where the residents could hide and ambush their pursuers. The inhabitants even created narrow tunnels up to 8km long to connect adjacent cities, in case their enemies managed to get past all their carefully laid traps.

I'm glad that I'd never had to spent months at a time in Derinkuyu, fearing for my life, but come to think of it, I do actually spend an inordinate amount of time underground. In fact, since I began working, I've spent a total of over 5 months of my life deep inside London's clay, as I take underground trains – the Tube – to work. Alongside millions of other people, packed into carriages like sardines, it's an uncomfortable reminder that, in my city, space is at a premium. The streets can't accommodate homes, offices, pedestrian paths, trains, trams, cars and cycles – not to mention water pipes, sewers, electricity and internet cables. And why should they? After all, we live in three dimensions and should use all of them, building up and down rather than simply sprawling sideways. The city beneath our feet is brimming with hidden engineering, but these arteries would not have been possible had it not been for the humble tunnel. In Derinkuyu, space was plentiful; tunnels provided safety. In London and many other metropolises, there is a lack of space, and tunnels provide the solution.

*

In the early 1800s, the only river crossing in the entire city was London Bridge – an immensely impractical and laborious situation in a metropolis that was spreading out rapidly on both sides of the Thames. The time taken to navigate the busy city, the wait to make the perilous and excruciatingly slow journey across the choking bridge, and the cost incurred in tolls were all sources of great frustration. In 1805 a company was set up to try to circumvent this by directly connecting the docks at Wapping and the factories at Rotherhithe.

Although the two points were only a tantalising 365m apart across the river, this distance was large enough to make building a bridge impractical – which meant that to get from one to the other, people and goods had to make an arduous 6.5km journey via London Bridge. Besides, putting a new bridge between the docks and factories would have stopped tall ships from reaching higher up the river, causing major problems for the thriving trade the city hosted. The only remaining option was to create a passage under the river. The problem was that canal builders, mining experts like Richard Trevithick, and other inventors had already tried to tunnel without success. The new company's efforts to bore a tunnel under the river were also unsuccessful, until an engineer came up with a solution inspired by a shipworm.

Marc Brunel was born in Normandy, France, in 1769. As a second son he was expected to become a priest, but he showed more interest in drawing and mathematics than in scripture, and entered the navy instead. Fleeing France in 1793 during the French Revolution, he went to America, where he eventually became Chief Engineer of the city of New York. He then moved to London in 1799, to try and persuade the Admiralty to purchase a new system he had invented for producing pulley blocks. He worked on various projects for the armed forces, developing apparatus for mass-producing soldiers' boots, and sawmill machinery at the Chatham and Woolwich dockyards. But he came to the attention of the Thames Tunnel Company (after vigorously lobbying its bosses) because of the tunnelling machinery he had invented.

Brunel carried a magnifying glass in his pocket. While working at Chatham Dockyard, he picked up a damaged piece of timber that had been removed from the hull of a warship, and scrutinised the actions of *Teredo navalis* (the naval shipworm) at close quarters. The worm had two razor-sharp, shell-like 'horns' on top of its head, and as it moved, wriggling and rotating its horns, the wood directly in its path was ground into a powder. The little shipworm ate the powdered wood and wriggled a few millimetres forwards into the space it had just created. The powdered wood travelled through the worm's digestive system and mixed with enzymes and chemicals in its body. The worm then excreted this mixture, creating a thin paste that lined the small tunnel left behind. When exposed to the air inside the cavity, the excretion hardened, shoring up the tunnel. Slowly but surely the worm moved forwards again and again, munching through the wood while creating a strong, lined passageway behind it.

Fully aware of the previous attempts to create a tunnel under the river, Brunel put his genius to work, and came up with a new plan. He realised he could succeed where everyone else had failed by adapting the process he had just observed. He would build his own shipworm: a machine that could tunnel forward and line the hole behind it. But his 'worm' would be made from iron. And it would be colossal.

Brunel's idea was that the device would have two blades, just like *Teredo navalis* – but that these would be twice as tall as a person. The blades would sit at one end of an iron cylinder lying on its side (and looking a little like a fan we might use in the summer to keep cool, but without the cage). A team of men would push the blades round so that they ate at the ground. Hydraulic jacks would push the cylinder forward. The soil which had been cut away by the blades would be transported backwards manually, like the shipworm excreting wood powder. As the cylinder moved forward, it would expose a ring of ground. To shore this up, bricklayers would lay bricks in a ring using quick-drying mortar to glue them together, creating a cylindrical shaft behind the blades, much like the worm's waste lining its tunnel. This process – turn fan, remove soil, lay bricks – would be repeated to gradually fashion a strong cylindrical tunnel.

Brunel's shipworm.

Having sorted out his worm, Brunel now had to find a suitable material for it to burrow into. Obviously, some substances are easier to dig into than others. Take dry sand, for instance. Fill a circular cake-tin with sand, then try to scoop out half of it to create a semicircle. You won't be able to, because the particles of sand simply collapse into the space you've just emptied. Similarly, if you try doing the same thing with very wet sand, the liquid nature of the material causes it to flow into and fill the space you're emptying. London sits on clay that's 50 million years old. If this clay is nicely compressed under layers of soil, and not too wet, it forms a fairly stable layer of ground. From an engineer's point of view this is good to work with, because you can slice into it quite easily, and it's unlikely to collapse. Put good clay – nicely compressed and not too wet – in a circular cake-tin and remove half of it, and you'll be left with a perfect semicircle of material. On the other hand, London's clay can vary considerably: it can be sandy, weak, watery and inconsistent. For Brunel's invention to work, he had to find good clay.

He hired two civil engineers to investigate in detail what the ground was made of. Paddling around in a boat, they plunged a 50mm-diameter iron pipe deep into the riverbed, then hauled it back out. They then studied the substances that had become trapped in it, looking to identify the different soils inside, and the thickness of each layer. After months of investigating they submitted their findings to Brunel, who decided that the ground was good enough for his plan to proceed without major problems. Before his shipworm could be let loose, however, he needed to burrow deep into the ground.

On 2 March 1825, the bells of St Mary's Church in Rotherhithe pealed as throngs of people made their way to Cow Court, ready to witness a very unusual sight. In the middle of the yard lay a huge iron ring 15m in diameter and weighing 25 tonnes. A brass band began to play as well-dressed ladies and gentlemen appeared, looking out of place in this rather squalid part of London. Amid cheers from the crowd, Marc Brunel arrived with his entire family, and was presented with a silver trowel, with which he laid the first brick on top of the iron ring. Brunel turned to his son, Isambard, who laid the second. Then followed speeches, drinking and toasts to the arts and sciences to mark the inauguration of the Thames Tunnel. But the joyful crowd had no idea just how much the sciences would be challenged in the months ahead.

The iron ring the crowds could see was like the sharp end of a cookie cutter. Two rings of brick separated by a layer of cement and rubble were laid on top of the iron ring, creating a cylindrical tower just under 13m high. On top of this the builders placed another iron ring, which was linked to the bottom one using iron rods sandwiched between the two brick walls. A steam engine was attached to the top of the 1,000 ton structure to pump away water and remove the excavated soil.

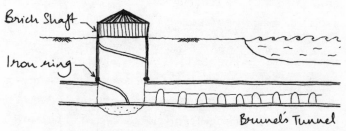

Brick Shaft

Iron ring

Brunel's Tunnel

Tunnelling under the river Thames, London.

To use a cookie cutter, we apply the strength in our arm muscles to push it down into the dough. But Brunel's idea was to allow his brick cutter to sink into the ground under its own weight: it was so heavy that it would naturally move through the soft soil. Slowly but surely, the shaft began to sink a few centimetres a day. As it sank, diggers removed soil from the middle of the cylinder, much as you would remove dough from the middle of a cookie cutter.

After getting stuck once, the brick shaft arrived at its final destination. To create foundations, the diggers dug another 6m below the bottom iron ring. In this space, bricklayers filled in three sides of the shaft and the floor, leaving one face open to the ground. This is where Brunel's 'worm' would be deployed to burrow the tunnel.

While all this was happening, Brunel realised that – unlike a shipworm, which could easily turn its blades – humans didn't have enough strength to rotate the blades of his tunnelling machine. He couldn't think of a way to attach a steam engine to provide the power, so instead he came up with a new idea. His solution was to divide the device into smaller sections – 36, in fact – with a single person working in each. He called this enormous machine 'The Shield'.

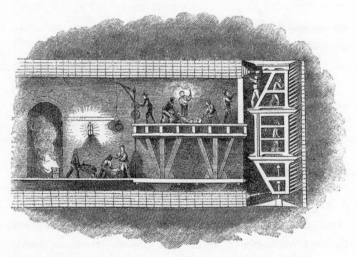

Working The Shield, the enormous machine used by Brunel and his men to excavate underground.

It had 12 iron frames, each 6.5m tall, 910mm wide and 1.8m deep. Each frame was divided into three 'cells', one on top of the other. The twelve frames were placed side by side to create a big grillage of 36 cells, each housing one worker, and these workers would operate The Shield. At either side of each man in his cell was a set of long rods, spaced at regular intervals from floor to ceiling. These held in place 15 or so planks of wood, stacked one above the other directly in front of the worker, and shoring up the ground in front of The Shield.

Operators in alternating frames (say frame numbers 1, 3, 5, 7, 9 and 11) worked simultaneously. Their task was to remove one wooden board by drawing back the two iron rods holding it in place, and dig out exactly 4.5 inches of earth and put the board at the rear of this new, slightly deeper cavity. They would then push the rods into place to support the board. The next step was to remove the subsequent plank and repeat the process, continuing like that until all the wooden planks in all the 18 cells had been fixed into their new positions. Now that these miners had excavated the section of ground in front of them, jacks at the rear of The Shield propelled their cells forward by 4.5 inches.

At this stage, the odd-numbered frames would be 4.5 inches ahead of the even-numbered ones. It was now the turn of the workers in the even frames to go through the whole process of adjusting rods, removing boards, digging into the earth and repositioning the boards. When they had finished, the even frames were pushed forward. The entire shield had progressed by 4.5 inches – the exact distance needed to fit one layer of bricks.

Behind The Shield was another flurry of activity. 'Navvies' (as the labourers who built the canals, roads and railways were known, after the word 'navigator') removed the excavated soil in wheelbarrows. Bricklayers stood on wooden planks and carefully laid bricks in the 4.5 inch gaps created as The Shield moved forwards. They used pure Roman cement, which dried very quickly and was incredibly strong – so strong, in fact, that when Brunel tested it by building a block of bricks and dropping it from a height, the cement didn't crack. He even had his workmen attack the block of bricks with hammers and chisels; while the bricks cracked, the cement stood unyielding. Brunel then decided to use this cement throughout the tunnel, despite its great cost (remember that a lot of energy goes into producing pure cement powder, which can be lessened by adding aggregate).

I try to imagine what it must have been like working in the tunnel. Before I'm allowed to set foot on a construction site, I have to pass exams, be trained in health and safety, and put on protective clothing. I walk around doing my job without worrying that I might not leave alive. Conditions in the Victorian tunnel were starkly different: the smell of the workers' sweat, the tallow smoke and the gas fumes made breathing very difficult – workers often emerged from the tunnel with a ring of black deposit around their nostrils. Flammable gases trapped in the soil were suddenly released and, if lamps were inadvertently brought near them, could catch fire and explode. The air was damp and the temperature rose and fell by thirty degrees, sometimes in the space of a few hours. It was also incredibly noisy – bricklayers shouting for more bricks, iron rods clanging, wooden boards thudding and hobnailed boots echoing through the tunnel. Brunel himself became very ill from over-exhaustion, and was prescribed the only treatment that would work: being bled by leeches on his forehead.

Brunel's son, Isambard, who was only in his early twenties at the time, became indispensable on the project as the main engineer running the site. (Sophia, Brunel's elder daughter, was nicknamed 'Brunel in petticoats' by the industrialist Lord Armstrong because Marc Brunel, unconventionally, taught his daughter about engineering. When they were children, Sophia showed more aptitude than her brother in all things mathematical and technical – and in engineering – but it was her misfortune to be born at a time when women had no such career possibilities. She is the great engineer we never had.) But Isambard, like his father, was often taken ill. And things were getting worse: the soil conditions were unexpectedly deteriorating, and funds were running out. At one point the whole operation was shut down and the tunnel bricked shut with The Shield inside it. It took six years for the Brunels to convince the Treasury to put more money into the project. The company directors meddled with Brunel's methods, refusing to obtain equipment he wanted to make the work safer, and pressuring him to work faster despite the risks. The biggest problem, however, was the flooding. The 'good' clay that Marc had been hoping to tunnel through was not consistent, and sometimes it disappeared completely, especially as the workers dug directly below the river.

The Thames was basically a huge sewer; all of London's waste (and many of the city's corpses) were deposited into it. The soil at the base of the river was very wet and of terrible quality, and the tunnel was being dug only a few feet below the river, right into this base. As The Shield moved forward, digging away at the ground, the soil was often displaced more than it should have been. There was also a weak point in the riverbed between The Shield and the brick tunnel, and if the soil was particularly bad it simply collapsed, sending river water coursing through the passageway.

The first time this happened, Isambard fixed the problem by contacting the East India Company and borrowing a diving bell (a chamber containing a couple of people that could be lowered underwater). In it he went to the bottom of the river, found the leak, and laid a bed of iron rods across the gap, with bags of clay piled on top to seal the hole. Once the water had been pumped away, the digging work could restart.

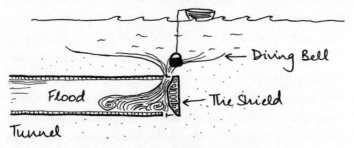

Innundation of the tunnel and the use of a diving bell to seal the breach.

This, though, was only the first of four major floods in which many men died. Isambard himself only narrowly escaped drowning, suffering his first (but not last) haemorrhage as a result, and being forced to leave the site for a few months' convalescence.

Despite the setbacks, however, in 1843, after 19 years' work, the tunnel was finished. Penny-paying pedestrians descended the spiral staircase in the shaft to the tunnel, which in its finished form was spectacular. A line of pillars down the centre supported immense brick arches. Gas lamps lit the passageway and an Italian organ powered by a steam engine played music. Hawkers sold refreshments and souvenirs from little alcoves in the brick walls. In 1852 the first Thames Tunnel Fancy Fair was held, featuring artists, fire-eaters, Indian dancers and Chinese singers.

But only a decade later, as the railways entered everyday life, the tunnel had fallen into disrepute. People no longer wanted to walk through its damp interior, choosing instead to take the flashy new trains. The tunnel became seedy and desolate, the haunt of drunks. In 1865 it was handed over to the East London Railway Company, and by 1869 rail tracks had been installed on the floor and steam trains began chugging through. Today, the London Overground line runs through it. The Rotherhithe shaft, which Marc Brunel managed to excavate so imaginatively, was recently opened to the public and has become a popular tourist attraction. Enter the stumpy circular tower and you find yourself in a cavernous underground chamber containing the remains of spiral staircases, and blotchy, scarred and weathered walls with mysterious black pipes feeding into and out of them. It's an incredibly atmospheric backdrop to the concerts and theatre performances that take place there.

Taking nearly 20 years to build, and then becoming obsolete just over 20 years after it was finished, the Thames Tunnel might not seem like a success. But thanks to Marc Brunel's imaginative engineering, we gained access to the underground parts of our cities. The London Underground – the first underground train network in the world – was made possible because of the work of Marc and Isambard Brunel, who showed us how to build structures in very fluid soil.

*

To dig their tunnels, the engineers building Crossrail (London's new train line) have been using a modern version of Marc Brunel's first and unsuccessful idea. Brunel couldn't get enough power to rotate giant blades, but electricity has made this simple for us. Instead of a manually operated machine, we use 'tunnel boring machines' (TBMs) – which are, of course, anything but boring.

Each of Crossrail's TBMs – described as 'giant underground factories on wheels' – is as long as 14 London buses end-to-end. The front has a huge circular cutter that spins, eating into the ground in front of it. An intricate jacking system pushes the machine forwards. Conveyor belts transport the excavated soil to the back of the TBM and out of the tunnel. A laser guidance system makes sure that the tunnel stays on course. Behind the TBM, a complex array of arm-like devices fix concrete rings in a circle (steel could also be used) to create the tunnel lining.

There's an endearing tunnelling tradition which proclaims that the TBMs must be named – with female names – before work can start. Crossrail ran a competition to name its TBMs in pairs, since the machines work in twos, radiating in opposite directions, starting from a point. One pair is named after the monarchs of the great railway ages: Victoria and Elizabeth. Another after Olympic athletes: Jessica and Ellie; another after the women who wrote the first computer program and drew the beloved London A–Z maps: Ada and Phyllis. Perhaps most fitting of all, though, are the names of the final two TBMs: Mary and Sophia, after the wives of the great tunnel builders themselves, Isambard and Marc Brunel.

PURE

It thrills me to see tourists taking pictures of buildings in a city, because it means that they love engineering – even if they don't realise it. They admire and respond to the ambition and the imagination that have gone into the design – curved canopies, tall silhouettes and unique facades are carefully selected, framed and frozen in time as the dramatic backdrop to countless photographs taken on phones mounted on selfie sticks. This architectural drama is the romantic side of engineering, and not to be underestimated. Nevertheless engineering is ultimately a response to very practical considerations; often it is less immediately exciting things like soil, materials or the law that are the driving force. A building or bridge might look spectacular; in fact, much of what shapes it can be decidedly unaesthetic.

One of the most influential of these considerations is water, which is such a fundamental requirement for humans that we can't survive much longer than three days without it. The structures I design are skeletons: until they have water, they are merely uninhabitable shells. I work with other engineers (mechanical, electrical, public health) to make provisions for the skeleton to support its circulatory system: creating pathways through it and making sure that its foundations, core walls and floors are strong enough to carry the weight of pumps and pipes. It's only when the arteries of water come to life that we create a building fit for the living.

But even though our planet is called 'the Blue Planet' because of the amount of water it contains, the shimmering, salty swathes of sea that cover most of the Earth's surface are not potable. We humans need easily accessible fresh water if we are to survive. But here's the problem: we don't actually have much of this. If all the water on our planet was represented as an area the size of a soccer pitch, then the freshwater lakes on the planet's surface would be the equivalent of the cushion I have on my sofa, while the surface rivers would fit inside the coaster I use under my tea.

Finding water is hard enough – and that's why many of our ancient towns were founded on the banks of a river – but as they grew into cities, as fields growing crops became vast, and as we migrated to live further and further from water sources, *moving* water became a challenge. It's no wonder, then, that in ancient times humans developed extremely inventive ways to track down and transport fresh water. Even today, engineers work hard to create solutions for this technically challenging process, and in parts of the world it is still a huge hurdle to be surmounted.

*

Like many others of the times, the ancients in Persia struggled to find fresh water. In the centre of Iran there is a large, dry, arid plateau that only receives a tiny amount of rain – less than 300mm – each year. As you fly over the country, desert stretches out below you, bleached of colour by the relentless glare of the sun. Occasionally, though, near small villages and towns, or even in seemingly uninhabited patches of the desert itself, you'll notice 'holes' in the sand. From a vantage point high in the sky, they look like the little crab holes that pepper the beach in Mumbai where I grew up. (I used to sit and stare at them for ages, waiting and hoping for a scuttling creature to appear.) But these holes are neatly arranged in straight lines, and are in fact much larger. Thankfully, they weren't the work of some giant crab, but were dug by humans, over the past 2,700 years. And throughout that time they have been essential to the survival of the people that live there.

These holes are part of the *kariz*, as it's known in Persian (or *qanat,* in Arabic): the system used by the ancient people of Persia to bring their life force – water – from below the ground.

To see how they were built, let's transport ourselves to the desert of two and a half millennia ago. The *muqanni* or worker looks near a hillside or slope for signs of the presence of water – a fan of deposited soils, perhaps, or a change in the type of vegetation. At a promising location, he takes a spade and digs a cylindrical well just over half a metre in diameter. To move the dirt he uses a windlass to haul a leather bucket full of soil up and down. Under the blazing sun, he keeps at it, hoping to find damp soil – a possible sign that the water table is close. Sometimes, he goes down as far as his tools will let him, but he doesn't find anything. At other times, he finds water hiding very deep, more than 200m down. Once in a while, he need only dig down 20m before he finds moisture. That's on a good day.

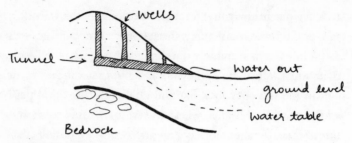

An ingenious kariz.

But the *muqanni*'s work has only just started: it's still possible that all he has found is a tiny bit of water that will quickly run out. He needs to make sure that his discovery is the real thing. So he leaves his bucket in the new shaft and, over the next few days, checks how much water, if any, has collected in it each morning. If he wakes up every day to a full bucket, he knows he has struck gold – or, rather, something even more valuable: he has found the face of the *aquifer* (an underground layer of permeable rock that contains water). He and his fellow *muqanni* then dig wells, one after the other, in a straight line down the slope of the hill.

Using a plumb line to measure depth, the *muqanni* dig each of these wells slightly deeper than the previous one. It may seem strange to dig a line of wells like this, but here is where the ingenuity of the *muqanni* lies: their village contains 20,000 people, and trekking up the hillside, drawing water and carrying it back would be a laborious task. Of course, this is done in many places around the world, but here the terrain – the hilliness and type of soil – means the *muqanni* can make the villagers' lives easier.

The wells finished, the workers start to dig a tunnel horizontally from the base of one well to the base of the next, creating a conduit about 1m wide and 1.5m high – just big enough for them to walk through so they can build the next phase.

This tunnel slopes gently, joining up the bottoms of the wells, and will bring the water out of the mountain. The slope of the tunnel is important: if it is too steep, the stream of water will be too strong and fast, eroding the soil and eventually causing it to collapse. If, on the other hand, the slope is too gradual, water will not flow easily, and will stagnate.

The *muqanni* light an oil lamp and place it at the mouth of the tunnel. And as they march into the mountain, they watch the flame so they can make sure they're working in a straight line. Noxious fumes may emerge from the ground to suffocate them, so the oil lamp not only acts as a beacon but as a kind of warning light: if the flame burns steady and bright, there's enough oxygen around. If it burns a different colour or goes out, it shows there are other gases present. There are other hazards too. Loose or crumbly soil could cause the tunnel to collapse, so where required the *muqanni* make hoops of baked clay and push them into the tunnel. The hoops act like two arches joined together: the weight of the loose soil pushes on to the hoops and puts them into compression. Clay is strong in compression, so the hoops reinforce the tunnel and stop it caving in.

There is a final hazard to be broached when the workers reach the head well (the first well, with its base at the face of the aquifer). They have to break through the aquifer very carefully, otherwise a jet of water might burst through and drown them.

Managing all this safely depends on the *muqanni*'s experience being passed from generation to generation: the techniques used to build *kariz* today haven't changed a great deal since ancient times.

The length of the conduits varies hugely, from 1km to over 40km. Some produce continuous water while others are seasonal. To maintain the system, the *muqanni* use the extra wells they dug. The frequent build-ups of silt and debris can be removed using the windlass to lower buckets into the wells. With regular repairs they can last a very long time.

There are said to be over 35,000 *kariz* in Iran – networks of hundreds of thousands of underground conduits all built by manual labour and still providing an important source of water. The city of Gonabad houses the oldest and largest known example in the country. It is 2,700 years old and its 45km conduit provides water for 40,000 people. The main well is deeper than The Shard is tall.

*

Digging down to an aquifer was one strategy the ancients had for supplying their citizens with water. But with water sources, terrain and tools differing across civilisations and eras, other ingenious solutions were invented, including many we still use today. By the end of the eighth century BC, the two canals providing water for Assyria's capital city, Nineveh, were no longer adequate to serve the burgeoning population. King Sennacherib (who reigned between 705–681 BC) – had previously used his engineering skills to dig canals through Babylon to flood and destroy it. Now, he was forced to find an additional source of water and channel it to Nineveh. He started nearly 50km away, at the watershed of the River

Atrush. From here he constructed a canal to the headwaters of the River Tebitu to increase the amount of water the Tebitu received. The river had earlier been dammed to create the reservoir that had supplied most of Nineveh's water. This extra water would flow to his city through the two existing canals, increasing its supply.

There was, however, one problem. To get from the river to the canals that led to Nineveh, Sennacherib's new conduit had to cross a small valley and, without a water pump, there was no way to push water up the far slope. Undeterred, Sennacherib conceived a structure that could carry water across the valley – what we know as an *aqueduct*. We think of the Romans as the foremost engineers of aqueducts, but the Assyrian king's edifice predates their efforts by several hundred years, making it one of the oldest such structures in the world. You can still see its remains at Jerwan in northern Iraq.

Technically, the word 'aqueduct' refers to any artificial channel used to transport water from one place to another: it can be a canal, a bridge, a tunnel, a siphon (a pressurised pipe), or any combination of these systems. The Nineveh aqueduct bridge was the greatest construction of Sennacherib, a master builder who also created much of Nineveh's civic architecture, including the legendary 'Palace Without a Rival'; he may even have been responsible for the Hanging Gardens of Babylon. Over two million cubes of stone went into the aqueduct's construction, each about half a metre wide. The end result was 27m long and 15m wide, made from pointed *corbelled* arches (a curved shape supported by projecting pieces of stone) that were over 9m high. A channel on top of the bridge allowed water to travel across the valley. The channel was lined with a layer of concrete to prevent the water from leaking away.

Load travels down not around.

A corbelled arch.

Incredibly, the new canal and aqueduct bridge were completed in only 16 months in 690 BC. When the structure was nearly complete, Sennacherib sent two priests to the upper end of the canal to perform religious rites. Before the allocated time for the ceremony, however, the gate holding back the water suddenly opened, releasing the river into the channel. The engineers and priests were terrified of the reaction this might provoke from the king, as Nature had defied his wishes. But the king decided this was actually a good omen, because the gods themselves were so impatient to see his great work completed that they had caused the gates to fail. He went to the head of the canal to inspect the damage, had it repaired, and rewarded his engineers and workmen with brightly coloured cloths, golden rings and daggers.

*

Finding and transporting water are two of the engineer's big challenges. But once you've got it, you have to know what to do with it: the third, equally important challenge, is storing it, ready for use. The Romans, who took aqueduct engineering to an impressively sophisticated level, came up with suitably ambitious storage solutions, such as the Basilica Cistern, situated in – or, rather, under – the centre of Istanbul in Turkey.

Basilica Cistern, Istanbul.

The Romans didn't invent the cistern: since at least the fourth millennium BC people in the Levant region (modern-day Syria, Jordan, Israel and Lebanon) had been building structures to hold water. Cisterns might seem like simple things to make, but in truth the biggest ones are impressive feats of engineering.

The Basilica Cistern, for example, has immense walls – up to 4 m thick – to resist the pressure from multiple gallons of stored water. To stop water leaking out, the Romans carefully sealed the walls with a coating of lime plaster about 10mm to 20mm thick. Since the roof of the cistern supported a public square, it had to be strong enough to support the weight of buildings, roads and pedestrians above.

When I visited Istanbul, the sun had pushed the thermometer to a stifling 35° Celsius, and I was grateful to descend the old stone steps into the cool air of the cistern's vast underground space. Uplighters emitted an orange-red glow and soothing music played in the background from speakers I couldn't see. I stepped onto raised wooden planks built recently to allow tourists to walk around. Below me, there was a pool of crystal-clear water a few inches deep in which grey, ghostly carp silently swam. I stood watching them, until I was jolted out of my daze by drops of water falling on my head and arms.

I looked up to see a roof made from beautiful red Roman bricks – the flat kind – with thick layers of mortar between them. Large arches spanned between numerous columns to create a grillage. Between these arches stood quadripartite vaults (domes which are divided into quadrants by four ribs). The breathtaking structure was held up by 12 rows of 28 columns, 9m high, all made from marble and arranged in a regular grid pattern. The tops of the columns varied – some had classical Greek and Roman designs on them; others were plain and bare – they had been salvaged from temples or other ruined structures. Some of the columns had split over time and were strapped together with flat pieces of black iron. A couple had the head of the Greek Gorgon Medusa carved at their base, the venomous snakes of her hair curled menacingly around her face. Her gaze was said to turn people instantly to stone, but here one of the carved heads lay upside-down while the other was on its side – a haphazard arrangement that somehow negated the deadly effect of her gaze. One column, known as the peacock column, was engraved with a curious pattern of circles and lines: these represent the tearful eyes of hens, and apparently the column was built as a homage to the hundreds of slaves that died during the cistern's construction.

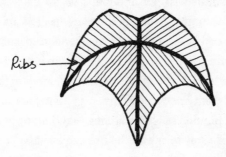

A quadripartite arch.

The Basilica Cistern was built by Emperor Justinian in AD 532. Lying beneath the Stoa Basilica, the large public square on the first hill of what was then called Constantinople (after the Emperor Constantine, who in AD 324 made the city the capital of the Roman Empire), it was capable of holding 32 Olympic-sized swimming pools' worth of water. The cistern received its water via an aqueduct that was connected to natural springs near the region of Marmara. It serviced the Great Palace, the residence of the Roman emperors, until they moved away, and it was subsequently forgotten about. In 1545, a scholar called Petrus Gyllius was talking to local residents as part of his research into Byzantine antiquities. After a little persuasion and coaxing, he discovered they had a mysterious secret – they could lower buckets through holes in their basement floors and miraculously haul up fresh, clean water. Sometimes, they even found fish swimming in their buckets. They had no idea why or how this happened – they were just glad to have a source of clear water (and sometimes even food), and until Gyllius came along, they had kept the secret to themselves. Gyllius realised that their homes must be above one of the famed Roman cisterns, investigated further, and found it.

I for one am glad he did – the place has a dramatic magic of its own, and has captured the imaginations of many people, including the thousands of tourists who have visited since it was refurbished and reopened in 1987. And, of course, the director of *From Russia with Love*, who filmed James Bond and Kerim Bey punting stealthily among the columns in sharp grey suits, on their way to spy on the Russian embassy.

*

It's incredible that something as big and impressive as the Basilica Cistern could simply be forgotten. It's incredible, too, how cavalier the Romans appeared in their attitude to water. Many historians believe that the rainwater they received was enough to live on, and that the aqueducts were for their baths and fountains. It seems extraordinary to perform such ambitious feats of engineering just for luxury and indulgence, particularly because in many parts of the world, then as now, water was in short supply and it took every ounce of an engineer's ingenuity to make it count.

In 2015 I visited Singapore to stay with a friend in her flat on the fourteenth floor of a tower block with wonderful views over the city. I checked with her that the tap water was safe to drink (of course it was) and that she had hot water available for a shower after my long flight. She warned me not to waste water, to turn off the shower when I was soaping myself and to make sure no water was dripping when I had finished.

I was impressed at her efforts to preserve water and be eco-friendly, but a longer conversation we had after my shower made me realise why this was. From a young age, it had been drilled into her by her parents, her school and her college that water is a precious resource not to be wasted. This is because Singapore has no natural aquifers or lakes. There are a few rivers that have been dammed to create reservoirs, but the country basically has no natural sources of water. Throughout its history, whether under British rule or as an independent nation, supplying its inhabitants with enough water has been a constant challenge.

The earliest sources of water in Singapore were streams and wells, which served the country adequately when the population was a mere 1,000. But after 1819, when Sir Stamford Raffles made the country part of the British Empire, the numbers greatly increased. By the 1860s, 80,000 people were on the island, and the rulers began building reservoirs to store water. In 1927, an agreement was reached with neighbouring country Malaysia, enabling the Singaporeans to rent land in Johor, from where they could pipe untreated water from the Johor River. In a reciprocal arrangement, another pipe from Singapore to Johor enabled the islanders to return some water once it had been treated. During the invasion and capture of the island by the Japanese in the Battle of Singapore (in 1942), the pipes were destroyed, leaving the people with enough water for just two weeks. 'While there's water, we fight on,' declared the region's commanding officer, Lieutenant-General Arthur Percival – but on 16 February he was forced to surrender.

This dire situation stayed in the minds of the people long after the Japanese left – to be replaced once more by the British – until 1963, when the country became, briefly, part of the Malaysian federation. So when Singapore gained full independence on 9 August 1965, water self-sufficiency was one of the government's top priorities.

In 1961 and 1962, Malaysia signed agreements to supply water to Singapore, one of which expired in 2011; the other is set to expire in 2061. For Singaporeans, it's a vulnerable position to be in, particularly in our water-dependent, high-consumption modern world, and I imagine they are concerned about their autonomy, given that they depend heavily on a neighbour for such a fundamental resource. If, for example, the whole area were to experience a drought, Singapore might end up at the mercy of another country. So for Singapore, water is as fundamental to its national interests as medicine or spies are to others.

As a result, Singapore is busy engineering a solution to its somewhat precarious situation. The Public Utilities Board (PUB) has developed a strategy called 'Four National Taps'. This refers to the four sources of water it will harness as efficiently as possible to provide a high degree of self-sufficiency for the country.

The first National Tap is rainwater. Singapore's location and exposure means it receives over 2m of rain every year. To conserve it effectively, engineers have created water catchments: areas of land where rainwater is collected rather than being allowed to drain away into the sea. A network of canals and basins has been built to trap the rain and channel it into dammed streams or reservoirs for storage. This has involved a massive clean-up operation, as over time many of the country's streams had become polluted by discharge from homes and businesses. So the PUB relocated polluting businesses and set about legally protecting the water stores from contamination. Rainwater is now being collected and stored in two-thirds of the island's land area. A few streams remain to be dammed – mainly those close to the sea, which have slightly salty water (which wouldn't be usable without some treatment). But once the engineers have finished, a massive 90 per cent of the land will be used, making Singapore the only place in the world that collects and conserves virtually all of its rainwater.

The second National Tap is water from Malaysia, which Singapore will continue to import until the agreement runs out. The third National Tap is recycled or reclaimed water. Although the practice of recycling waste water is not new – Los Angeles and other parts of California have been doing it since the 1930s – it is still far from commonplace.

Singapore first started thinking about recycling waste water in the 1970s, when the appropriate technology was still too costly and relatively unreliable. Eventually, however, it improved to the point where the project became viable, so now waste water is collected from homes, restaurants and industry and subjected to a three-stage purification process, using the latest in membrane engineering.

The first stage is *microfiltration*, during which the water is passed through a semi-permeable membrane. This is typically made from synthetic organic polymers such as polyvinylidene fluoride, which allow certain atoms or molecules to pass through but not others, and filter out solids, bacteria, viruses and protozoan cysts. Essentially, the membranes are microscopic versions of a colander, holding onto solids but allowing liquid to drain through. The water that escapes still has dissolved salts and organic molecules in it, so the second stage of recycling is designed to remove these, using a process called *reverse osmosis*.

Osmosis is the movement of a solvent (a substance that can dissolve others – the most common example is water) from a less concentrated solution to a more concentrated one, until the two concentrations are equal. It is an important part of our natural world – the means by which plant roots absorb water from the soil, for example, and by which our kidneys extract minerals such as urea from our blood. You can see the process in action for yourself, using an egg, vinegar, and treacle or corn syrup. First, soak the egg in vinegar for a couple of days, to dissolve the calcium in the shell and leave what is in effect an osmotic membrane. Then put the egg in treacle or corn oil.

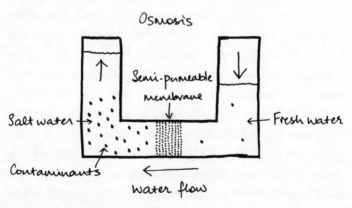

The process of osmosis.

Over the next few hours wrinkles will appear in the surface of the egg as water leaves through the membrane, dehydrating the egg in the process. Remove the shrivelled egg and put it in fresh water, and you'll see the process reverse, as water goes into the egg via the membrane, plumping it back up.

Osmosis happens naturally: fresh water filters through to mix with salty water easily. But if you want to produce more fresh water, you need to use pressure to 'push' the salty water through the membrane, which blocks the salt, bacteria and other dissolved matter. The pressure you apply needs to be bigger than the natural osmotic pressure, so you can force fresh water molecules through the semi-permeable membrane. This is reverse osmosis.

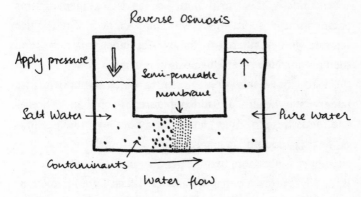

The process of reverse osmosis.

Reverse osmosis can remove up to 99 per cent of dissolved salts and other contaminants. So while the water coming out of this process is already of a high quality, there might be a few bacteria or protozoa still in it. As a backup, the water is disinfected using ultraviolet light to kill off any remaining microorganisms, and then it is ready to be distributed.

In 2003, after years of testing, NEWater – which is what the recycled water is called – was introduced to the public. During the parade of Singapore's 37th National Day, the Prime Minister, Goh Chok Tong, the founding Prime Minister, Lee Kuan Yew, and the thousands of people attending all opened a bottle of NEWater and sipped it while the cameras rolled. No one got ill. In fact, NEWater is used mostly in industrial estates and fabrication plants that require water of an even higher quality than drinking water. NEWater has passed over 100,000 tests and actually surpasses the World Health Organization's requirements for water that's fit for human consumption – even if its origins make you squirm.

And finally, the fourth National Tap is seawater. In 2005 Singapore opened its first desalination plant in Tuas, where seawater is first filtered to remove the largest particles, and then put through reverse osmosis in much the same way as for NEWater. The result is pure water, to which the minerals we need to stay healthy are added, before it's supplied to homes and industries. The Tuas plant can produce 30 million gallons of water (130,000m³) a day. The third and fourth National Taps already produce more than 50 per cent of the country's needs. By 2060 it's projected that the scheme will account for about 85 per cent – a spectacular and potentially life-saving transformation brought about by clever planning and engineering.

*

That Singapore collects most of its rainwater for reuse and is planning for long-term water sustainability demonstrates how engineering can solve critical, real-world problems. It's an age-old challenge, involving the most basic and essential of molecules, but one which is now being addressed using some of the most advanced technology available. As time goes on and our global population increases – and with it the demand for water – engineers and scientists across the planet will have to confront the escalating challenges of locating this precious liquid, creating new pathways to channel it, and enhancing the science to purify it.

Otherwise, we will not survive.

CLEAN

My visit to Japan in 2007 was one of the most memorable and inspiring trips I've been on. My mum and I wandered the streets of Tokyo marvelling at the vending machines that dispensed eggs, fruit, ramen and even puppies, and we ate at sushi restaurants where enthusiastic chefs and waiters shouted out everyone's orders in a harmonious chorus.

I was also intrigued by the toilets, which played music, and which featured buttons that lit up, and cleaning sprays that automatically sanitised, making a normally mundane act an exciting affair. In my experimentation, I did press a few buttons and regretted it pretty quickly – but, hey, I felt cleaner afterwards, if a little violated. When we left Tokyo for more remote locations, we encountered much more basic squat toilets: it was a stark contrast – but nothing compared to medieval Japan.

Long before the Tokugawa shogun regime (1603–1868) was established in the country, solid human waste – euphemistically known as 'night soil' – was being traded. It was loaded onto ships that sailed all around Japan, distributing it. Unsurprisingly, the ships carried a rancid stench with them, and people complained about these fetid vessels being docked alongside ships carrying tea. Magistrates, however, decided that the trade was vital, and that people would just have to deal with the stench.

Trading human faeces was important because of the particular challenges this small island nation faced. Because of its topography, Japan had little land for growing crops, yet the population was booming and increasing food production was essential. So the land available for agriculture had to be used intensively to produce enough food, with more than one harvest per year. This meant that the natural nutrients of the soil were rapidly becoming depleted. Traditionally, the Japanese had turned animal waste into fertiliser to replenish the soil, but there weren't many animals on the island, so the inhabitants had to look elsewhere for a solution. They found the answer in their own sanitation: the burgeoning population created a lot of waste. So the Tokugawa shoguns decided to make a virtue out of necessity by removing the waste to ships, and then trading it with farmers looking to boost their crops.

The turd trade was soon big business. During the early years of the Tokugawa shogunate, the country began to depend on one of the biggest cities at the time, Osaka, for fertiliser. Boats laden with vegetables and fruits would arrive in the city and exchange their produce for its citizens' night soil. However, the value of the night soil quickly increased (inflation affects faeces too, apparently) and vegetables were no longer enough to pay for such a valuable commodity: by the early eighteenth century, people were buying it with silver. Laws came into force stating that the rights to faecal matter produced by the occupants of a dwelling belonged to the landlord, though they were generous enough to assign the rights of urine to the tenants themselves. The price of faecal matter from 20 households a year amounted to the same as the cost of grain one person would eat annually. Night soil was by now an integral part of the housing market: the more tenants that landlords had, the more waste they could collect, so the cheaper the rent.

Eventually farmers, villagers and city guilds were all fighting over rights to buy night soil. By the mid-eighteenth century, lawmakers in Osaka assigned ownership and monopoly rights to officially recognised guilds and associations that would determine a fair price. Even then, the high prices crippled the poorer farmers, and people risked harsh jail terms by turning to theft.

Night soil collection may have become a cause of conflict, but it had some unexpected benefits, too. Because waste was collected so obsessively and carefully, the water sources people used to collect drinking water were less likely to be contaminated. Other cultural practices helped: the Japanese drank most of their water in the form of tea – boiling the water got rid of many disease-causing microbes. And those who followed the ritual practices of Shinto, had strong views about sources of uncleanliness – blood, death, illness – and 'purified' themselves if they came into contact with anything unclean. All this meant that life in Japan in the mid-seventeenth to mid-nineteenth centuries was more sanitised and hygienic than in many countries in the West, and the Japanese suffered lower mortality rates as a result.

The twentieth century was different. With the constantly growing population and the Second World War causing devastation (not least in economic terms), the good quality of life people had enjoyed deteriorated. In 1985 only about a third of communities had modern sewage systems – a lag caused mainly by the success of the pre-modern methods for dealing with waste. In the 1980s the sewage network was modernised, and nowadays the Japanese are famous for their advanced toilets, in extreme contrast to the night soil trade that flourished not so long ago.

Whether in modern times or the distant past, the way in which a city deals with its waste has been an indicator of how successful and enterprising it is. Almost every home in the cities of Harappa and Mohenjo-daro in the Indus Valley Civilisation (around 2600 BC) was connected to a water supply and had a flushing toilet. In our densely packed post-industrial cities, efficient waste disposal has always been of vital importance. As Florence Nightingale (whose hygiene initiatives revolutionised Victorian hospitals and homes) acknowledged in an 1870 Indian Sanitary Report: 'The true key to sanitary progress in cities is, water supply and sewerage.' Those of us lucky enough to have a great sanitation system rarely give a second thought to where our poo goes once the toilet flushes. Those that don't, on the other hand, are all too aware of the disease and death that festering waste can bring. It might be a subject that makes most of us squeamish, but as the population of our planet rockets, adequate sanitation is becoming increasingly important.

*

'The trouble is,' said Karl, 'no one gives a shit about poo.' He stormed off.

At the time, I was working on the design of a small apartment building near Oxford Street in central London. While I was busy arranging columns around the car-parking bays and the swimming pool in the basement, my drainage-engineer friend Karl was working out how much waste water would be produced by showers, sinks and toilets inside the building, and by rainfall out-side. Once he had calculated the amount of flow per hour, he had to make sure there were enough pipes to convey it into London's sewage system. From historical records, we knew there was a large sewer adjacent to our structure, but we didn't know exactly how big and full it was, or whether it was in reasonable condition. We wanted to know if we could use it to discharge waste from our structure, but also if digging the basement near this sewer would damage it. Karl had written to a survey company to gather infor-mation about the pipe so he could complete his design.

One day, Karl turned up with a DVD and, without much explanation, asked me to feed it into the computer and press play. Almost at once I shrieked and scrambled to turn the thing off. Among my colleagues, in the middle of my office, on my computer screen – which suddenly seemed enormous – were being displayed the results of the sewage survey. I hit the stop button and told Karl I wouldn't watch it – and that's when he told me off and strode away.

Chastened, I sat down, took a deep breath and clicked play. The film was shot by a small camera mounted on a robot on wheels being driven through a sewer wirelessly by someone standing safely on the ground. The brick walls were a deep red colour and looked pretty clean despite the unappetising con-tents that had flowed through them for the past 150 years. The sewer was surprisingly large – I could have walked through it without crouching – and fashioned into a distorted oval a bit like an egg standing on its narrow tip. This shape helps waste to flow easily – in times of low flow, the speed of the effluent is high since it's in the lowest and narrowest portion of the sewer; at high flows, the larger crown creates space.

The amazement I felt watching this robot moving through a landmark piece of engineering easily overrode any feelings of queasiness I had at seeing what lay at its bottom. Over the next week, Karl and I (having quickly put our faecal fracas behind us) studied the film in detail, and decided that the nearby sewer was intact and in good condition, so waste from the new building could be discharged into it. (We couldn't simply dump it all in when it suited us because of the risk of overwhelming the sewer. So, like in many buildings in London, we created an 'attenuation tank' in the basement where the waste is stored and released into the pipe at an acceptable rate.) It was an exciting moment for me: I was creating a real physical link to the pioneering engineering work done by Joseph Bazalgette more than a century ago, when he envisioned and built a vast network of sewers under the capital. At that time London sorely needed such a system, for in the early nineteenth century, living in London was a very disgusting experience.

*

Originally, the plains of London were served by a number of tributaries that provided plentiful water and fish on their path to the River Thames. But as the population of the city increased considerably in the mid-thirteenth century, the quality of the water deteriorated. Things got worse until, eventually, the tributaries were nothing more than open sewers and dumping grounds for animal and even human corpses. By the fifteenth century, 'water carriers' made a livelihood for themselves collecting water from wells in two barrels tied to a stick across their shoulders, but the rivers were in such a state that even going upstream didn't help. The water the citizens of London were drinking was contaminated with their own waste, and dead bodies.

The city also housed 200,000 cesspits – cylindrical pits, often lined on the inside with bricks in an attempt to keep them watertight, about 1m in diameter and twice as deep, with a sealed base and a lid at the top. Their purpose was to store human waste: people would take the chamber-pots they had used to relieve themselves, and empty them into these tanks. It was then the job of the 'nightmen', 'rakers' or 'gong-farmers' ('gong' was apparently a medieval term for a latrine) to periodically clean these out, carrying the waste in buckets to fields. This was better than having the waste in the streets, but it was still decidedly unhygienic, given that the fields weren't especially far from central London. Cleaning the pits was obviously unpleasant work, but it was also dangerous – spare a thought for Richard the Raker, who in 1326 fell into a cesspit and was asphyxiated and drowned in a putrid mix of urine and faeces.

Attempts by the Commissioner of Sewers to pass Acts to build new sewers in the 1840s remained inadequate. Introducing 'water closets' (or the modern-style water-flushed toilets) only made the situation worse: the leaky cesspits had been barely adequate to hold concentrated waste, but now litres of water were being emptied into them, flooding them. In 1850, to try and overcome this, the pits were banned, but as a result the sewers (which were only designed to take away surface water from rain) became completely overwhelmed. All waste – human and other – ended up in the Thames, which was still used by people for washing, cooking, and drinking.

The vile mixture of waste and water in London led to severe and devastating cholera epidemics. They usually struck in late summer or autumn, and half of the people who contracted the disease died. The outbreak in 1831–2 killed more than 6,000 people; it was followed by two more major outbreaks in 1848–9 (just over 14,000 dead) and 1853–4 (another 10,000 fatalities). The common belief at the time was that cholera was airborne and that you contracted it by inhaling a poisonous 'miasma'. But during the 1854 outbreak Doctor John Snow (1813–1858) monitored the health of people drawing water from a contaminated pump in Soho, and collected evidence that this was not the case: cholera was, in fact, spread by contaminated drinking water.

Thomas McLean's etching 'Monster Soup commonly called Thames Water' of 1828 was a grotesque satire on the city's water supply.

That London's waste was ruining the capital became particularly apparent during the unusually hot summer of 1858, which warmed up the banned but festering cesspits and the sewage-filled Thames and its tributaries, so that the city smelled even more horribly pungent than usual. And so the 'Great Stink' (as it was called) began. It became so unpleasant that people soaked their curtains in a lime chloride mixture to try and hide the stench. The smells were so noxious that ministers working in the House of Commons, and the lawyers at Lincoln's Inn, were unable to work, and they made plans to abandon the city.

The only upside of all this was that, having been affected by the awful conditions first-hand, the government finally became determined to get rid of the stench and the cholera that came with it. In 1859, after years of rejecting plans from engineers to solve the sewage problem in London, officials finally approved the works proposed by Joseph Bazalgette.

Bazalgette was described as having an indifferent temperament but a pleasant and genial smile. He was considerably below average height but his long nose, keen grey eyes and black eyebrows gave him the impression of being a powerful man. He was born in Enfield on the outskirts of London in 1819 and pursued a career as a civil engineer. A taxing stint working on the rapid expansion of the railways led to a nervous breakdown in 1847, after which he became a surveyor for the Metropolitan Commission of Sewers, tasked with solving the problem of drainage in London. He was later appointed to the Metropolitan Board of Works, whose job it was to devise a solution to London's problems with waste disposal.

Bazalgette's plan made use of the Thames' old tributaries, which were now basically sewers, and which had been diverted to flow along brick culverts or channels. The diversions helped to satisfy the demand for more housing: restricting the rivers to narrow culverts allowed people to build homes close to the edge of the water. These culverts were often buried under roads, freeing up even more space. Their highest point was away from the river, and they flowed down in a north–south direction until they reached the Thames (which flowed west to east), where they deposited their putrid water.

Joseph Bazalgette decided he would intercept these culverts and their horrible contents. He did this at various points, creating a web of new sewers that sat below the old culverts. Inside these old culverts, to partially block the water flow, he constructed weirs (a form of water barrier) that were half as tall as the height of the culvert. Then, in front of these weirs, he bored holes through the floor so that most of the waste water would be redirected into his new sewers below. Hold up your left hand with your fingers spread out, then put your right hand below it at right angles to the left, and you've got a good representation of Bazalgette's system. Your left hand is the series of old tributaries flowing through their culverts; your right is Bazalgette's new sewers.

North of the river, Bazalgette installed sewers below the culverts at three points. The first was far north, where the culverts were relatively high (for those familiar with London, this branch runs from Upper Holloway through Stamford Hill and Hackney down towards Stratford). About halfway between this 'high-level' sewer and the river, he installed a 'mid-level' sewer, running from Bayswater, below now world-famous shopping area Oxford Street, and Old Street. This collected more of the wastewater as it hit the weirs and poured down through the holes in the base of each culvert. Finally, very close to the river, he put in a 'low-level' sewer to capture the remaining water. South of the river he did something similar but used only a high-level sewer (running from Balham through Clapham, Camberwell and New Cross to Woolwich) and a low-level one (going from Wandsworth through Battersea, Walworth and on to New Cross). This was because there were fewer people living there, and the extent of the city south of the river was less than in the north. In total, this system, end-to-end, would have measured 160km.

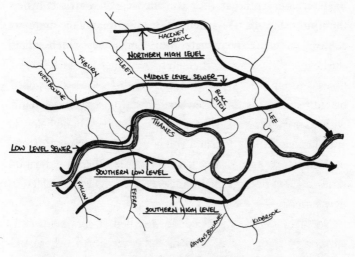

Bazalgette's main sewer network that reached across London.

The Victoria, Albert and Chelsea embankments in London are all products of his work. These contain the low-level sewers that run alongside the River Thames. Just as engineers before him had restricted the width of the tributaries of the Thames by putting them in culverts, Bazalgette narrowed the mighty river itself with these embankments. His new underground routes not only housed the new sewers, but also created space for the first underground railway: the London Tube.

When designing the five main sewer pipes and the hundreds of offshoots, to calculate the size required Bazalgette made a generous allowance for the amount of waste produced by every one of the 2 million inhabitants of the city. Then, figuring that these sewer constructions would only be done once, he doubled the size. His five sewers were at their highest point where they began in the west, and they sloped two feet for every mile as they travelled east towards two new pumping stations. Designed by Bazalgette and the architect Charles Henry Driver, these were Crossness (which served

The interior of the Victorian pumping station with its decorative ironwork at Crossness Sewage Treatment Works, Erith, London.

the two southern sewers) and Abbey Mills (which served the three northern ones). Solid, imposing and cathedral-like, both pumping stations are masterpieces of late-Victorian architecture. The real surprise is the interior of Crossness, in which the vast pump machinery is surrounded by gleaming brass and extravagantly ornate, colourfully painted wrought ironwork. In fact, the stations have appeared on screen several times, notably in films such as *Batman Begins* and *Sherlock Holmes*.

Having travelled through the sewers and reached the pumping stations, the waste had to be lifted back up to a level high enough that it could flow naturally to large sewage storage tanks further east. North of the river, the waste was stored at Beckton, while south of the river it was stored in a tank next to the Crossness pumping station. The reason the waste needed to be raised was so it could flow under gravity into the Thames when the river flowed out towards the sea on its ebb tide. At this time, the contents were still being dumped into the river untreated.

Bazalgette was told to make sure his tanks were far enough east so that, in the worst case, if they *had* to be emptied during an incoming tide because they were full, the back-flowing sewage wouldn't come as far west as Westminster – the ministers didn't want a repeat of the smells they experienced in 1858. In fact, by narrowing the river, Bazalgette inadvertently caused its tidal range to go further back than before, so occasionally the smells did get quite fusty.

Although the idea behind Bazalgette's sewage system was in essence quite simple, executing it was not, as building the new sewers meant digging up London's roads. It must have been an incredibly invasive and complex piece of work, digging down to the right level, constructing egg-shaped brick sewers and the connections to the culverts, then filling the hole and redoing the roads. But it was worth it, because life in the capital slowly began to get better.

The quality of water in central London improved dramatically. Bazalgette's sewers (2,100 kilometers of them, made up of more than 300 million bricks) were finally completed in 1875. By that time, the ravages of cholera in London were a thing of the past, in large part due to Bazalgette's practical, efficient and imaginative piece of engineering.

*

Bazalgette took the sewage from central London and deposited it outside the city into the River Thames, which ultimately took it out to sea. The waste was not treated, so the system basically moved the disease-causing elements from a populous area to a deserted one. If this sounds to you like a somewhat old-fashioned approach, then it may come as a surprise to learn that we use exactly the same system today.

Nowadays, in new waste systems, the water collected from rainwater goes ideally into pipes that are separate from those that pick up sewage from homes and offices and industrial waste from factories and restaurants. The idea is that the rainwater, which is not polluted, can be discharged into seas or rivers, while the sewage and industrial waste are taken to treatment plants.

At the plants, the polluted waste is broken down into its base chemicals, using a series of physical, chemical and biological processes. 'Physical' could mean filtration: passing water through membranes to remove impurities. 'Chemical' is the addition of substances to the waste, which react with it to break it down. 'Biological' is a similar process, but using bacteria to break down the waste. The aim is to create 'treated effluent' – an environmentally safe fluid for disposal – or 'sludge', solid waste that can also be disposed of or used as an agricultural fertiliser.

That, at least, is the theory. In practice, it rarely works like that. Shockingly, estimates by UN-Habitat (an agency monitoring places where people live) state that, globally, 90 per cent of waste water is released into the environment untreated or after only primary treatment. And at the moment, London is no exception. This is because Bazalgette's sewers are 'combined sewers', which means that they carry everything – rain, sewage, industrial effluence. Bazalgette was incredibly forward-thinking in designing the sewers for the waste of 4 million people (twice the population of Victorian London) plus rainwater. Now, though, the population in London is 8 million and we are still using this system, which is nearly 150 years old. The reason it still works most of the time is because the sewers are big enough to cope with the 1.25 billion kilograms of poo they receive each year. But since the system is working at relatively full capacity, it can't cope with rain as well, so even if it rains just 2mm in a day (which is a common occurrence in dear, damp London), these combined sewers fill up and overflow.

Dotted along the sides of the River Thames are 57 pipes that discharge this overflowing waste directly into the river. You can see the exit point for one of these at Battersea, where there is a large, reinforced iron door set into the river bank; there's another under Vauxhall Bridge. The Vauxhall one alone currently releases 280,000 tonnes of waste a year. Some of these exit points were built in Bazalgette's times, while others were added later. In 2014, excess flow had to be discharged into the river more than once a week, equating to 62 million tonnes of untreated sewage released into the Thames every year. That's the equivalent in weight of more than 8,500 blue whales plunging into the river *every week*. If we do nothing, that will almost double by 2020. Such statistics are liable to make anyone breathe uneasily. Fortunately, however, between now and 2023 a huge project to deal with this problem will be under way, under the feet of unsuspecting Londoners: the Thames Tideway Tunnel.

I made an appointment to see Phil, one of the directors of the project to create a new 'bowel' for the capital. We settled down in a large canteen to chat about urine and faeces – or, more specifically, how they will now be removed in a more modern manner.

'Our scheme is an extension of Bazalgette's legacy,' Phil explained. 'One that, I believe, he would have done himself if London's population had grown to such extents during his lifetime.' The premise of the project is simple: 150 years ago, Bazalgette intercepted the decaying tributaries. Now, the Tideway Tunnel will intercept Bazalgette's sewers: instead of the wastewater from his sewers overflowing into the river, it will overflow into a new network of tunnels.

The scale of the project is impressive. At 21 sites around the city – including one at the Vauxhall discharge point – new vertical, cylindrical shafts will be dug up to 60m deep to collect the excess sewage. Most of these will be built at the edge of the river. The first step is to build a large *cofferdam*, a watertight enclosure where the construction site can be set up. Within this area, a new shaft will be installed close to the existing sewage discharge point. Chambers will then be built to connect the existing discharge point to the shaft. So, instead of flowing into the river, the sewage will flow through the chambers into the new shaft. As Phil pointed out, while it's fine providing a new system, it's also extremely important that it's invisible, both to sight and smell (I pictured living next to a large toilet). Acres of public gardens and parks will be developed on top of these shafts. So in a few years' time, you'll be sitting on a bench by the river sipping your cappuccino, surrounded by grass and trees, while literally tonnes of sewage per second pour from Bazalgette's sewers into the shaft below you. When the waste reaches the bottom of the vertical shaft, a pipe will carry it through to the new tunnel.

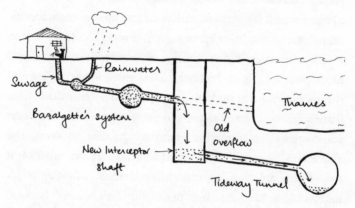

Intercepting sewage via the planned Tideway Tunnel; the future of the sewage system within London.

This main artery is 7.2m in diameter: large enough to contain three double-decker buses side by side. It starts at Acton in West London and falls 1m for every 790m that it runs east. By the time it reaches the pumping station at Abbey Mills, the tunnel is as deep as a 20-storey building is tall. From Abbey Mills, the sewage is pumped to the Beckton sewage treatment works.

The majority of this tunnel runs below the River Thames in central London, which is a really interesting engineering strategy. It's an excellent idea to do this, because running new infrastructure under a busy city is difficult at the best of times. But London in particular has a large underground tunnel network and thousands of buildings with deep foundations. By running the tunnel below the water itself, it passes under only 1,300 buildings (which might seem a lot until you consider how many more it would have been if the tunnel had run under land instead). It also goes below 75 bridges and 43 tunnels, including the Tube tunnels, as it burrows under the city.

The ground itself poses another huge challenge. Since the tunnel runs across the city, and slopes downwards from west to east, it encounters different soil at different places. At the start in Acton it goes through clay, which is prone to expanding and contracting. In the middle section, through central London, it runs through mixed sands and gravels, which are problematic materials to tunnel through because they move around and aren't cohesive. Finally, in the east, in Tower Hamlets it runs through a chalk layer with big chunks of flint in it. It's impossible to predict where all this flint will be, and, because it's difficult to cut through, it can cause delays as the tunnel boring machines (TBMs) struggle to slice their way through the ground. The tunnel needs to be strong, especially at the junction between two different types of soil, because one type of soil might be much more cohesive or drier than the other, and apply different forces to the tunnel as it expands and contracts. Five TBMs will work at the same time in different parts of the city, moving in different directions, to form tunnels that will eventually join up to create the 'super sewer'.

The aim of this mind-boggling project is to bring down the number of discharges into the river from 60 a year to four, reducing the amount of wastewater from 62 million tonnes to 2.4 million tonnes a year. I asked Phil why the discharges couldn't be completely stopped, and he explained that these four discharges would happen only when there is very heavy rain: during such storms, the sewage is diluted considerably as the storm water mixes with the waste, so the discharge into the river is not toxic. The oxygen levels in the river would not really be affected by these diluted overflows because of the natural biological processes in the water that maintain its ecosystem. To reduce the discharges to zero, the Tideway Tunnel would have had to be twice as big.

Engineers often need to compromise in this way: the ideal solution is not always the most practical one. Ideally, we would have separate sewer pipes for rainwater and for waste, but this would mean having to more or less shut down London and dig up all the streets to put in a brand-new system. Even more ideally, we wouldn't discharge into the Thames at all, but this could actually be worse for the environment. Creating a tunnel of the necessary size would mean removing twice as much soil from the ground, resulting in a much longer construction process with bigger machines and much more energy. This method would also reduce the amount of water in the river itself, as the flow of the natural tributaries would be completely shut off.

The Thames Tideway Tunnel project will clearly have a momentous effect on the quality of the river: no longer will swimmers and rowers have to worry about sloshing through human waste. But what made me even happier was when Phil pointed out that the project will incorporate new treatment plants. We've come full circle from Bazalgette's solution, and are adding another network of shafts and tunnels to his system to meet the needs of the modern city. But this time we will be decontaminating our waste so we don't contaminate our seas.

Today, we pay homage to Bazalgette for having the skill and imagination to create a sewer system we are still able to use, 150 years later. Hopefully, the current expansion of the system will serve us just as long and, in a century's time, city dwellers will be thanking us for giving London a new bowel.

That's probably enough about poo.

IDOL

When I walk into a room for a meeting, I'm often the only woman there. Sometimes I keep a tally – 11 men and me, 17 men and me. The most, I think, was 21 men and me. Surrounded by men, I conduct my business, bemused when one of them swears, then looks sheepish and apologises directly to me (they've clearly never seen me driving my car in heavy traffic). I have opened countless work-related letters addressed to 'Mr Agrawal' – after all, if you can't tell my gender from my name, you have a greater than 90 per cent chance of being right if you go for male. Because, much to my frustration, I am in a minority in my profession.

Working in a man's world can be challenging in all sorts of ways, sometimes comical, other times trying. It's hard to keep a straight face and conduct professional conversations about finite element modelling or soil strength profiles when I'm in a site office surrounded by pictures of naked women. On one occasion a builder asked me if I wanted my picture taken in my 'costume', in other words, the hard hat and hi-vis jacket I wear regularly for all the site visits that are part of my job. I've heard stories from other women in the industry about how they've been (illegally) asked in job interviews when they plan to get married and have children.

Thankfully, these are mostly occasional occurrences. And ultimately I love what I do and believe that anyone can succeed in my field with persistence and resilience. I acknowledge that being in a minority can even have advantages – people tend to remember me after a meeting because I've spoken knowledgeably about concrete and cranes while wearing a chic dress and shoes. And it has provided some unusual opportunities to be a spokesperson for engineering, such as fashion and make-up shoots.

My engineering idol: Emily Warren Roebling.

I admire many engineers – I've talked about many of them in this book – but Emily Warren Roebling holds a special place in my heart. She understood technical concepts as well as any of the male engineers churned out by universities that wouldn't admit women, yet she never trained as an engineer: she simply learned when she had to. Her brilliant communication skills earned her the respect not only of labourers on site but also of the highest-ranking politicians of the time. What's more, pioneering innovations in engineering were implemented under her watchful eye.

Being in a minority and working in construction has its difficulties in the twenty-first century, but Emily did all this at a time when most believed that womens' brains were not even capable of understanding the complex mathematics and engineering she mastered. Her masterpiece, the Brooklyn Bridge, remains one of the most iconic symbols of New York.

*

From a very early age it was clear that Emily was extremely intelligent and had a keen interest in science. Despite a 14-year age gap, she shared a very close relationship with her oldest brother, Gouverneur K. Warren. He entered West Point military academy at 16 and then joined the Corps of Topographical Engineers, surveying for future railroads and mapping areas west of the Mississippi. He went on to fight with distinction in the American Civil War (a statue to him stands at the entrance to Prospect Park in Brooklyn). Warren was Emily's hero. When their father died he assumed responsibility for his family, encouraging Emily's interest in science and arranging for her to be enrolled in the Georgetown Visitation Convent, a preparatory school for women. There, she further explored her passionate interest in science, history and geography, as well as becoming an accomplished horsewoman. In 1864, during the American Civil War, Warren was posted far away, but Emily made the arduous journey to visit him and, during her stay, met Warren's friend and fellow soldier Washington Roebling. Contrary to her usually balanced and sensible nature, she fell in love at first sight. Six weeks later, he bought her a diamond ring.

Throughout the rest of the war, Emily wrote long affectionate letters full of details of her life. But Washington destroyed them soon after he read them, saying that the letters made their separation much more painful to him. Emily, on the other hand, saved everything he ever wrote to her, and in less than a year she had more than 100 letters containing all his thoughts, fears and affections. While he was away fighting in the war she visited his family, and they took a great liking to her. Finally, after 11 months of correspondence, Emily and Washington Roebling were married on 18 January 1865, and Emily stepped seamlessly and gracefully into the role of a typical Victorian housewife: tending to house and family in the shadows of her husband.

Washington's father, German-born John Augustus Roebling, was an accomplished engineer, and Washington planned to follow in his father's footsteps. In 1867 John sent Washington to Europe to study building methods, one of which was inspired by the Romans.

*

The relatively light and small structures that the Romans built in the early years of their empire didn't really need foundations because the ground was strong enough to carry them. But as they mastered the techniques of construction, their structures increased in size and weight, and the Romans learned that foundations were a crucial part of the design of their creations, which would otherwise move or sink. It was relatively easy to build foundations on land by digging out the soft earth and laying strong stone or concrete on the firmer, deeper layers of earth. Doing the same in a river, though, was – as you might imagine – more complicated. But being the inventors they were, the Romans came up with a solution.

They sometimes supported their structures by driving piles made from logs into the ground. They inserted the piles using *piledrivers*: machines made from inclined pieces of timber connected together in a pyramid shape, and about two storeys tall. Pulleys and ropes attached to the apex of the pyramid allowed men or animals to raise a heavy weight. A wooden log was pushed into the ground as deep as could be achieved manually, and then the rope holding the weight was released, dropping it on top of the log and pushing it further. The process was repeated until the log was completely submerged.

To create foundations in water, the Roman engineers started by using piledrivers to install wooden piles into the riverbed in a ring around the position of their intended foundation. They inserted two concentric rings of piles, and then packed the space between them with clay to seal it. The water within the ring was bailed out, leaving a dry area in which they could work. This sort of construction is called a *cofferdam*. It's a technique still used today (in the Thames Tideway Tunnel, for example, as we saw in the previous chapter), but using large steel piles shaped like circular tubes or trapezoids.

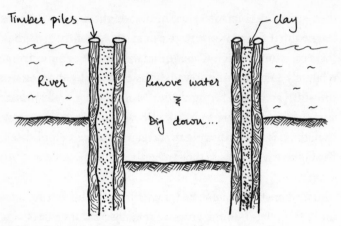

Building foundations in water the Roman way.

Inside the dry cofferdam, Roman workers dug out mud until either they hit rock, or the cofferdam started leaking. On top of the strong ground they built a pier of stone or concrete in layers. (Using their special pozzolanic cement they were able to make solid concrete even in damp and soggy environments.) Once the pier was built, they piled rocks against the base to stabilise it further, then put mud back into the hole to its original level. The base of the pier or column and the pile of rocks were buried in the riverbed. The timber piles were then removed and water flooded back in. The workers continued to build the pier as high as it needed to be to support the bridge structure above.

*

The Roman cofferdams worked in places where the water was not too deep. But Washington Roebling wanted to know how to go deeper. Driving piles wouldn't work because they would be really tall and wouldn't be able to resist the push of the water. So he studied *caissons*.

A caisson is a chamber with a watertight top and an open base, which penetrates into the mud of the sea- or riverbed. (You can picture this by pushing an upside-down tumbler into a pot of water that has sand at the base: the tumbler rim pushes into the sand while the sealed top keeps water from coming in.) One chute from the surface provides access to workers so they can descend into the chamber, and another is the passage through which they can take materials in and out. But if you want to really go deep into the water, there is another challenge. The further you descend, the greater the pressure of the water, and the harder it pushes against the walls of the caisson.

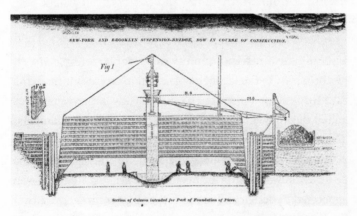

The immense caisson used during the construction of the Brooklyn Bridge.

To counteract this pressure, you can use a *pneumatic caisson*. These are 'normal' caissons with an added feature: compressed air is pumped into them. The pressurised air stops water from coming in and also balances the push of the water on the sides. An airlock gives workers access to the chamber. Engineers started to use these literally groundbreaking innovations to install foundations for bridges around the middle of the nineteenth century, and Washington Roebling was fascinated. He even considered using explosives in the confined space – a technique that, for obvious reasons, hadn't been tried before.

Emily began to assist her husband's research, studying caissons alongside him, and using the scientific methods she had learned at the Georgetown Visitation Convent to understand bridge engineering. Little did she realise at the time, that the dangers of working in the highly pressured environment of a caisson would eventually lead to a catastrophic change in their lives, one from which Emily and her husband would both emerge very different people.

*

Until the late nineteenth century there was no bridge connecting Brooklyn to the island of Manhattan, and although ferries shuttled back and forth across the East River, they were often halted during winter when the river froze over. So there was great pressure on the government to improve the situation. A bill was passed chartering the New York Bridge Company to do exactly that, and in 1865 John Augustus Roebling was appointed to design and make cost estimates for a bridge over the East River. They were to arrange for funds, which were to be split between the City of New York and the City of Brooklyn (which at the time were separate cities), along with private investors. Two years later, John Roebling began to lead the entire project.

The central section of the bridge he designed took the form of a suspension bridge, which has some similarities to the cable-stayed form I used for the Northumbria University footbridge: both employ tall towers to which cables are attached. And in both types the cables are always in tension, which holds up the deck. However, the two bridges differ in the way the tension force channels itself into the ground.

The journey of the forces in a cable-stayed bridge is direct. As the deck pulls down on the cables, putting them in tension, these cables, which are connected directly to the towers, compress the towers. In a suspension bridge, however, the weight of the deck pulls on cables that in turn pull downwards on *another* cable – a *parabolic cable* – which is suspended from the tall towers at each end. (*Parabolas* are curves with a particular shape – for the mathematically minded, if you plot a graph of $y = x^2$, you get a parabolic curve.) The parabolic cable is anchored to foundations on the opposite side of each of the towers. The parabolic cable exerts a downward force on each tower, putting them into compression and channelling the forces into the foundations. This is the difference between the two types of bridge: cable-stayed bridges don't have parabolic cables.

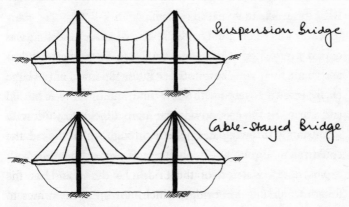

Suspension versus cable-stayed.

Work on the Brooklyn Bridge began in 1869, but disaster struck almost immediately. An accident on site left John Roebling with tetanus and he died a few weeks later, without even seeing the first stone of his spectacular structure laid.

Washington Roebling was the natural successor to his father, and took on the role of Chief Engineer on the project. To sink the piers for the bridge, he made use of the caissons that had caught his imagination in Europe. But his were larger than any that had been used before, and he was also going much deeper under water. With layers of heavy stone on the lid, he drove two huge chambers – each 50m wide by 30m long – into the river, one on the New York side and another on the Brooklyn side.

While this looked to be a reasonable engineering decision on paper, reality soon sank in. During the first month of excavation, progress was so slow that the engineers questioned whether they should give up and start again with a different approach. As columns of black smoke emanated from steam engines, and tar barrels, tools and stacks of stone and sand cluttered the site, reports began to surface from the workers about what it was like to be in the caisson.

It was incredibly loud in the restricted space, and shadows darted everywhere; the pressure affected the workers' pulses and made their voices faint. Every internal surface of the large chambers was covered with slimy mud, and the air was humid and warm. As the ground became more difficult to work with – constantly throwing up boulders through which the caisson couldn't cut through – Roebling began experimenting with explosives. He worried about the quality of the air and how his design would affect his workers, not knowing at the time that his own health would soon be ruined.

Over the next few months, having spent hours deep below the surface, Washington suffered exhaustion, temporary paralysis and deep pain in his joints and muscles. He had even hired a doctor to be on hand to supervise the condition of the men working in the Brooklyn caisson, which was deeper than the New York one. Without a full understanding of the health issues that he and his workers were facing, Washington shrugged off his symptoms and continued working. But even though the pain was temporary, the feeling of numbness in his extremities was not. He became a victim of *caisson disease*, in which nitrogen is released into the blood, causing acute pain (liable to make the sufferer double over, which is why the condition became known as 'the bends') and even paralysis or death. Now, of course, we understand the dangers of moving from high- to low-pressure environments too quickly – divers, for example, ascend at a rigorously controlled rate so the gas can be expelled. In 1870, however, caissons were a relatively new innovation and, although the engineers knew something about the dangers of working at depth, they had yet to determine the mechanism for avoiding injury.

Washington was in constant pain – in his stomach, his joints and his limbs – and severely depressed. Ravaged by headaches, he was losing his eyesight and was upset by the slightest noise. But only he had the knowledge and ability to oversee the project in his father's place. Nevertheless, Washington's physical condition made it impossible for him to be actively involved; even normal day-to-day tasks were now a struggle. His mental state left him loath to speak to anyone except Emily. It seemed as though all the years of design and planning that the Roeblings had put into the bridge, and all the personal sacrifice they had endured, were going to count for nothing. Emily, however, had spent a long time with her husband and father-in-law, hearing about bridge design and engineering, and even helping with the technical research. Slowly, she began to get involved. It was, however, a huge step. The idea of a woman being involved in the project, and perhaps even leading it, was unprecedented. Apart from the doubts and mistrust everyone would likely feel for Emily – from the builders on site to the investors – did she herself have the confidence and resolve to act as a liaison between her husband and the site, let alone take over the role of Chief Engineer?

With some background in science, but no detailed knowledge of bridge design, Emily began by taking extensive notes from her husband. She feared that he would not live to see the bridge completed. She then took over all correspondence on her husband's behalf, regularly writing to the offices of the company. With unwavering focus, she started to study complex mathematics and material engineering, learning about steel strength, cable analysis and construction; calculating catenary curves, and gaining a thorough grasp of the technical aspects of the project. Emily was determined to see her family's legacy built.

She soon realised that these skills alone were not enough for her to successfully lead the project: she had to communicate with the workers on site, and the powerful stakeholders. So she began visiting the site every day, instructing the labourers on their work and answering their questions. She supervised the build and relayed messages between her husband and the other engineers working on the project.

As Emily grew in confidence, she relied on Washington less and less. Her gut instincts guided her decisions and her blossoming skills helped her anticipate problems before they happened. Records of all work on site and responses to letters were diligently filed, and she tactfully represented her husband at meetings and social events. When bridge officials, labourers and contractors visited looking for her husband, she intercepted, answering their questions with authority and confidence. Most of them left satisfied, and many of them addressed all future correspondence to her – and in their minds she became the true authority. (At one point during the build, there were investigations into the honesty of some of the suppliers. In 1879, representatives of one of the contractors, the Edge Moor Iron Company, keen to allay suspicions, wrote a letter addressed to 'Mrs Washington A. Roebling' that made no mention of soliciting opinions from her husband.)

Yet Emily conducted her work in Washington's name. Rumours circulated that she was the actual Chief Engineer and the real force behind the bridge. News outlets made oblique references to her: the *New York Star* commented archly about 'a clever lady, whose style and calligraphy are already familiar in the office of the Brooklyn Bridge'. During the entire period of construction, the Roeblings kept their home life strictly private – no magazines or newspapers were permitted to interview them.

Despite Emily's careful handling of the project, problems started to proliferate. Costs mounted. Twenty men died from accidents and the caisson disease. Washington's health showed no signs of improvement. The so-called 'Miller Suit' had been filed. Warehouse owner Abraham Miller sued the cities in charge of the bridge, demanding that they remove the structure in its entirety; claiming that it would divert trade to Philadelphia; challenging the cities' ability to fund the project; and presenting a number of shipmasters, shipbuilders and engineers who would testify against the safety of the steel cables used in the bridge. Only the determined efforts of Senator Henry Murphy, a long-time supporter of Washington's father, led to the suit finally being settled. Even the Roeblings didn't escape accusation – it was claimed they had conducted questionable transactions with steel manufacturers, and they were investigated for accepting bribes, before eventually being cleared. The board of trustees overseeing the build changed and political dogfights broke out between new and old members. And then, in 1879, the Tay Bridge in Scotland – one of the biggest and most famous bridges in the world at the time – collapsed in a gale, killing 75 people. A headline in the *New York Herald* wondered: 'Will the Tay Disaster Be Repeated Between New York and Brooklyn?'

In 1882, despite Emily's skilful and assured work on behalf of her husband, the Mayor of Brooklyn decided to replace Washington Roebling as Chief Engineer on the basis of physical incapacity. He passed a motion with the Board of Trustees to dismiss Roebling, calling for a vote at their subsequent meeting. After much argument, political wrangling, and reporting in the press, they gathered, debated, and cast their ballots.

By a majority of just three votes, the men agreed to let Washington Roebling continue running the project until its completion. Nearly half a lifetime later, when Roebling was asked what part Emily had played in building the bridge, he answered 'her remarkable talent as a peacemaker' among the divisive personalities involved in the bridge's construction. I like to think of her as the polished negotiator: patiently listening to every side of the numerous arguments, offering tactful words of caution to the men, and smoothing difficulties in a highly-charged political atmosphere. Emily was clearly instrumental in ensuring her family's legacy remained intact.

Before the bridge was opened to the public, one final test had to be conducted: checking the effect of a trotting horse on the structure. Even at that time, the dangers of resonance – movement caused by users of a bridge – were well understood, so precautions were taken to establish that the bridge was stable and safe for different modes of transport. Carrying a live rooster as a symbol of victory, Emily was the first person to ride across the bridge in a horse-drawn carriage.

A few weeks later, on 24 May 1883, she was given the honour of accompanying President Chester Arthur's procession as he officially opened the bridge, while her husband watched on proudly through a telescope from his room. The day – which came to be known as 'The People's Day' – was declared an official holiday in Brooklyn. Fifty thousand residents spilled into the streets, celebrating and hoping to catch a glimpse of their President and their new bridge. Numerous speeches revered the bridge as a 'wonder of science', and an 'astounding exhibition of the power of man to change the face of nature'. Or, in this case, the power of woman. During the ceremonies, Abram Hewitt, one of Washington Roebling's competitors, stated: 'The name of Emily Warren Roebling will . . . be inseparably associated with all that is admirable in human nature and all that is wonderful in the constructive world of art', and called the bridge '. . . an everlasting monument to the self-sacrificing devotion of a woman and of her capacity for that higher education from which she has been too long disbarred'.

The official opening ceremony of the Brooklyn Bridge.

Today, on one of the towers supporting the bridge there is a bronze plaque dedicated to the memory of Emily, her husband and her father-in-law. Placed there by the Brooklyn Engineers' Club, it reads:

'The Builders of the Bridge
Dedicated to the memory of
EMILY WARREN ROEBLING
1843 – 1903
Whose faith and courage helped her stricken husband
COL. WASHINGTON A. ROEBLING. C.E.
1837 – 1926
Complete the construction of this bridge
From the plans of his father
JOHN A. ROEBLING. C.E.
1806 – 1869
Who gave his life to the bridge
"Back of every great work we can find
The self-sacrificing devotion of a woman"'

Emily Warren Roebling was technically brilliant and liked by just about everyone she ever worked with. She was held in high esteem and shown great respect by the forces behind the bridge, regardless of their role or aspiration for the project. That she, as a woman, could traverse every social circle, and was welcomed by politicians, engineers and workers, her opinions heeded and instructions followed, was in itself proof of her exceptional skills, in an age when a woman's presence on a construction site was unheard of.

*The commemorative plaque to the Roebling family on
the Brooklyn Bridge.*

As a young structural engineer at a similar age to Emily when
she was working on the bridge, I am well aware of the challenges
and pressures involved in constructing a key architectural
landmark in a major world city. But I came to my greatest
engineering challenges after years of structured technical
training, experience, guidance and support – gaining my
chartered engineer's qualification on the way. Emily did it without
any formal training; she was not even a qualified engineer. Tragic
circumstances forced her into a situation in which she never
expected to find herself, yet she excelled and triumphed. This
was not just any bridge – its 486m span made it by far the longest
bridge of its time. It was the first to use steel wires for suspension
cables, and the first to employ caissons of such enormous size,
and explosives within them. It was a pioneering structure that
has persisted to this day.

In my research I have been surprised to see the disparity in the way Emily's contribution is acknowledged by commentators. In some places, she is highlighted as the true force behind the project. In other sources, there is absolutely no mention of her at all. But, compared to equivalent women of her time, her contribution has received some recognition at least. I am delighted that her name endures on the commemorative plaque. She is an inspiration to me because, despite the monumental challenges she faced, she delivered the most advanced bridge of its time, using every skill an engineer needs – technical knowledge, the ability to communicate with labourers and persuade stakeholders, and tenacity – at a time when women were expected to be silent and inconsequential.

BRIDGE

'*Flirtman* called again. Managed to get rid of him in only 3 minutes and 23 seconds.'

At a party, I had been introduced to a man who chattered away at me, altogether too suave and flirtatious for my liking – or rather, the type that considers himself suave but isn't really. Eventually I extricated myself and was careful to steer clear of him for the rest of the evening. But not quite careful enough – somehow we ended up swapping phone numbers.

Over the next few weeks, he called me a couple of times. The first time, my mum had just arrived from India, so I fobbed him off with a polite, 'Sorry, my mother has just arrived, I can't talk now.' The second time I got rid of him in just over three minutes, and proudly emailed a friend to tell her so.

But Flirtman – as he'd become known to me and my friend – was persistent. He called and emailed a few more times (the conversations began to extend past three minutes). Finally, I agreed to go on a date with him. It was then that I found out something unexpected about this young man – he was a complete geek. We talked about physics, programming, architecture, history; and I discovered that he spent hours reading

Wikipedia, and that his brain had an uncanny capacity for interesting but essentially useless facts. I left dinner hiding the little flutter I felt.

I can't think how it happened, but over the course of that dinner Flirtman spotted that I too am a bit of a nerd, and he developed a cunning strategy to get my attention. The morning after our first date, I opened my emails to see a message headed: 'Bridge of the Day no. 1'.

'An example of why you should do a proper damping analysis,' read the email: it was the Tacoma Narrows Bridge in Washington which collapsed dramatically in 1940 in a relatively light wind. Each morning after that I'd log on, still half asleep, and a grin would spread across my normally grumpy face as I saw that a new Bridge of the Day had appeared. In fact, for a whole week, he found and sent me a Wikipedia link and a picture of a bridge: one which had a funny story, a unique design, had suffered a catastrophic failure or just looked beautiful. Was I that obvious? Surely it couldn't be that simple to win me over . . .

Even though I still thought of the sender of the emails as slightly trying, I enjoyed his bridge stories, and learned about examples I hadn't even heard of. After a week of such offerings, I at least had to acknowledge that he'd pulled off a pretty good chat-up line. It's not every day you get serenaded by a series of bridges. And so, in honour of Flirtman, here is my version of Bridge of the Day. I've chosen five of my favourite examples from around the world – but unusual or obscure ones that, hopefully, you haven't heard of. Each bridge is made from different materials, ranging from silk to steel. I've chosen them from different periods in history, and they demonstrate different methods engineers had for building. One bridge moves because it's designed to, one is unintentionally bouncy, and one was made by an ancient king. Each has its own unique engineering feature – offering a glimpse into the thousands of creative ways that humans have crossed valleys and rivers through the ages.

No. 1: Old London Bridge

Old London Bridge: that was often falling down.

This isn't a bridge I've seen, because it was finally demolished in 1831. With its tumultuous history it holds an air of mystery for me: it's the legendary bridge – built thanks to the passion and perseverance of one person – that spanned the Thames for more than 600 years. What fascinates me above all is that for centuries it served the people of London faithfully – but, ultimately, badly. Despite its impressive longevity, Old London Bridge failed as a structure.

The Romans, as you might expect, were industrious and efficient bridge-builders. But after their western empire declined in the fourth and fifth centuries AD, very few bridges were built until the 1100s. At that point, the Church started to fund and construct a large number of bridges. Many of these had chapels where one could pray for safe passage, and contribute financially to its upkeep. There is a legend that Saint Bénézet (who was inspired by a vision to build the famous Pont d'Avignon) founded the Fratres Pontifices or 'Brothers of the Bridge', who built bridges wherever they were needed for religious or community purposes.

Spurred on by this development, Peter of Colechurch, curate of a small chapel in London, decided to raise funds to build a new bridge over the River Thames in 1176. He collected donations from the king, peasants, and everyone in between, in order to build the first stone bridge in London. Previously there had been a wooden bridge that had been variously destroyed by storm, fire, military strategy or simple neglect. Building this structure, however, would prove to be a big challenge for Peter, as it was the first time anyone had proposed a bridge with stone foundations in a tidal river. The Thames is not an easy stretch of water to bridge: it moves up and down by almost 5m, has a very muddy bed, and contains fast-flowing water, making it extremely difficult to build the foundations and piers to support the deck. Even getting materials to the site promised to be a struggle, as stone had to be bumpily transported over the poor-quality cobbled roads provided for travellers. Undaunted, Peter took on this mammoth task.

People in medieval London must have been dumbfounded by the elaborate construction of their first stone bridge. They would have heard the ear-splitting thuds of the piledrivers, mounted on barges, which slowly wound up a large weight, then dropped it to whack piles into the riverbed. They would then have seen artificial islands called *starlings* built on top of the piles. Each was shaped like a rowing boat, and was constructed by amassing stones and rocks of different sizes. The starlings – and the piers or columns that rose from them to support the deck of the bridge – were huge and irregular in size, ranging from 5m to 8m in width. The populace watched as carpenters attached wooden skeletons shaped like arches to the piers. These were centering, on top of which carved stone was placed (after it had been perilously lifted from barges) to create the arches. The people of London had to wait an entire year to see just one arch completed.

In 1209, 33 years later, the bridge – which was 280m long and nearly 8m wide – was completed, but Peter of Colechurch did not live to see it. He died after 29 years of service to the structure, and was buried in the crypt of its chapel.

The finished bridge was extremely crude. It had 19 arches of different shapes and sizes, made from randomly cut stone in the pointed Gothic style. Although the pointed arch inspired by Islamic architecture was all the rage in buildings and churches at the time, it was not an efficient shape to use for a bridge. Certainly, such arches allowed medieval churches to be taller than ever before – but the bridge didn't need to be tall, it needed to be the right height to link both sides of the river. A more traditional semicircular Roman arch would have been more appropriate, but it looks like the engineers were going for style over substance. At its centre was a drawbridge to allow tall ships to pass through, and each end was surmounted by a defensive gatehouse.

The River Thames rises and falls with the tides. By blocking nearly two-thirds of it, the overly broad starlings and piers of the bridge severely restricted the natural flow of the river. So, when the tide turned, the water was much higher on one side of the bridge than the other, because it couldn't flow past, and the choked water created deadly rapids. All but the most foolish sailors avoided passing under the bridge during those times, for fear that their boats would overturn, casting them into the river. But hundreds died. Maybe their lives would have been saved had they paid heed to the saying, inspired by the bridge, which cautioned that it was made 'for wise men to pass over, and for fools to pass under'.

To make matters worse, houses began appearing on the bridge. Now, I like the idea of living on a bridge – watching the river change as the day went on and enjoying spectacular sunsets would undoubtedly have been an uplifting experience. This has worked beautifully on the Ponte Vecchio in Florence in Italy, where the carefully planned and executed houses and shops create a feeling of peace and civic order. By contrast, the houses on London Bridge only added to the chaos.

Squashed between the carriageway and the edge of the struc-ture, numerous three- and four-storey houses and shops were built, until there were over a hundred such dwellings. Temporary stalls were set up in front where shopkeepers sold their wares. Public latrines overhung the sides of the structure, discharging waste directly into the river below. The bridge had not been designed for the weight of the buildings, and the buildings themselves were not safely separated from one another, creating a huge fire risk. The bridge really was an accident waiting to hap-pen. Most of the houses were destroyed by a fire in 1212, along with thousands of unfortunate people who had crowded onto the bridge to watch the flames take hold at one end – but then strong winds carried embers to the opposite end and started a new fire, trapping them in the middle. More than 3,000 bodies were found severely or partially burned, and many more were reduced to unidentifiable ashes. In 1381 and 1450, revolts and rebellions again laid waste to many parts of the bridge.

By the fifteenth century, the buildings on the bridge had doubled both in number and in height. These tall, overhanging structures created dark and dismal passages through which carts, wagons, cattle and pedestrians fought their way. At peak times it could take an hour to cross. Between the overloading of the bridge by the houses, the effects of fires, and the wearing away of the supporting piers by the rapids flowing between them, some portion of the structure was always crumbling and collapsing into the water.

In 1633 a third of the homes were destroyed by yet another fire, although this was perhaps a blessing in disguise, because it created a gap between houses on the bank and those on the bridge. This probably saved the structure from disaster in 1666, when the Great Fire of London couldn't spread across this void. It was, quite literally, a narrow escape, but it seems that the residents and shopkeepers didn't learn their lesson. Another fire in 1725 destroyed over 60 houses and two of the arches.

*

The houses were finally demolished in 1757, and the bridge survived past the turn of the century until 1832, when the new London Bridge (designed by civil engineer John Rennie) was constructed alongside it. But the original bridge is still firmly embedded in our culture – when I was little, my mother taught me the nursery rhyme inspired by its precarious history, singing, 'London Bridge is falling down, my fair lady' in her slightly accented, out-of-tune voice. It's a rare song about engineering. It teaches the future engineers among us about the perils of bad design before we can even walk.

No. 2: The Pontoon

THE BRIDGE OF BOATS OVER THE HELLESPONT, USED BY XERXES.

The Pontoon: bridging the sea with boats.

When we think of a bridge, we usually picture something high up in the air, neatly straddling the obstruction it needs to avoid. My second bridge, however, defies this image. Seeking revenge, the ancient Persian king Xerxes built an immense 'bridge' to cross nothing less than the sea. But instead of flying over the water, he used its buoyancy to create a unique bridge, known as a *pontoon*.

Xerxes' father, Darius I, was one of the greatest emperors in history, ruling unopposed from the steppes of central Asia to the tip of Anatolia. His empire was far larger than Alexander the Great's (and, under his successors, it grew larger still). Between 492 and 490 BC he decided that the tiny Greek city states must fall under his rule, and he marched to Marathon to battle an army from Athens and Plataea. His surprise defeat there marked the end of the first Persian invasion of Greece.

Darius had planned a second attempt but died before he could fulfil his plans. Xerxes never forgot the humiliation his father had faced at Marathon, and was determined to fulfil Darius' dream of bringing the Greek states under the heel of the Persian Empire. Xerxes spent years training soldiers, planning and accumulating supplies before he attacked, and once again, while most of the Greek states submitted to him, he faced resistence from the Athenians and the fierce warriors of Sparta.

A challenge arose in 480 BC when the Persian army needed to march into Thrace via the Hellespont, the strait (now known as the Dardanelles) that separates modern-day European and Asian Turkey. After the first attempt at a crossing failed when a violent storm destroyed the bridges the Phoenicians and Egyptians had built, Xerxes ordered that the waters be given 300 lashes for their insolence. And he had the engineers who built the two failed bridges beheaded.

The replacement engineers, presumably in a bid to save their necks, built a much more substantial structure. The Persians had to travel 1.5km across a deep strait – at the time a huge distance to span, and very difficult to do using the traditional bridging technique, which was to build foundations underwater on solid ground and then span material between supports. Instead, as Herodotus tells us in *The Histories*, they gathered 674 ships (a combination of *penteconters*, Greek ships with 50 oars, and *triremes*, low, flat ships with three banks of oars) and arranged them side by side in two lines. Laid above each row of boats were two cables of flax and four cables of papyrus. These extremely heavy cables tied the boats together, and created the base of the deck.

The engineers cut long planks of wood from tree trunks and laid them edge-to-edge on top of the taut cables. The planks were tied together and evenly covered with a layer of broken twigs and branches, after which soil was thrown on top and stamped down to create a surface that the army could walk on. The engineers also laid heavy anchors upstream and downstream of the bridge: those to the east stopped the boats being pushed down the strait by winds from the Black Sea, while the others resisted the winds from the west and the south. Fencing was installed along the sides of this wide walkway to prevent the horses from seeing the water and being spooked.

Once this bridge of boats was ready, Xerxes offered prayers for safe passage. He threw his cup, a golden bowl and a Persian sword into the straits, possibly as an offering to the sun, or possibly as a form of appeasement to the sea. The army then began to cross this monumental pontoon bridge en route to the Greeks at Thrace. It is said that it took seven days and seven nights for the Persians – including Xerxes' elite fighters known as 'The Immortals' – to cross from one side of the strait to the other.

Despite this feat of engineering, the military side of the story is less epic. Xerxes was defeated at the battles of Salamis and Plataea and, after losing large numbers of men to war or starvation, he retreated back to Persia. Although he managed to subjugate Nature, Xerxes couldn't do the same to the Greek people.

*

Floating or pontoon bridges are believed to have originated in China sometime between the eleventh and sixth centuries BC, when engineers used boats with boards on top to cross large rivers. Use of the pontoon bridge continued through ancient Roman and Greek times – a notorious example was supposedly assembled by Caligula so he could show off his clothes in parades. During the World Wars, soldiers often used this technique because it allowed them to assemble and dismantle a path across water quickly and efficiently. Floating bridges are a great option when water is deep, the span is long and time is short. But storms and currents in the water affect them badly: there are many examples (such as the Murrow and Hood Canal Bridges in the USA) which have failed in strong storms. If one of the boats fills up with water, it drags down the others, until the whole line sinks. Fortunately, engineers no longer face the same fate as those that once served Xerxes.

No. 3: The Falkirk Wheel

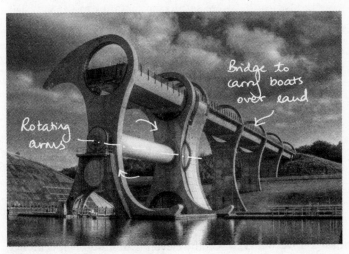

The Falkirk Wheel: a rotating bridge.

As it bounced up and down with the waves, and sideways with the currents, walking across Xerxes' boat-bridge would have been disconcerting. We don't like our structures to move perceptibly: it scares us into thinking that they are not safe. But what if a bridge is designed to rotate? Many bridges allow land vessels to cross water, but one of my favourite bridges enables water vessels to cross land.

The Celtic doubled-headed axe was a formidable weapon, with a blade on either side of its shaft so that, in battle, a brave warrior could swing it right or left with equally destructive results. Unlikely as it may seem, this menacing tool is the inspiration for one of the coolest and most unusual structures in the world – the Falkirk Wheel.

The low-lying canals of Scotland were once a flurry of activity. The Union Canal, opened in 1822, went from Falkirk to Edinburgh, as a way to bring coal into the capital and feed the new industries that were setting up factories in the city. The Forth and Clyde Canal (opened in 1790) served the same purpose for Glasgow, at that time a small town that was rapidly growing into the industrial heartland of Scotland. However, once the railway network began to develop in the 1840s, these canals, like so many others, became redundant, because it was quicker to transport minerals by train. The canals gradually fell into disrepair – by the 1930s they were in such a state that a portion of the canal system was filled in. A former transport artery was sealed off for good – or so it seemed.

At the end of the twentieth century, architects and engineers conspired to reopen the canals by creating a new waterway-based link between Glasgow and Edinburgh, specifically between the Forth and Clyde Canal and the Union Canal.

Making the 200-year-old waterways usable again offered environmental and economic advantages to the communities that lie along them. But doing this presented some technical challenges, foremost among them a large steep slope that had to be crossed. The traditional way canal-builders dealt with a slope was by means of a *lock*. Between the lower and higher sections of canal, they constructed a long, narrow, tall-sided chamber with a door (or pair of doors) at each end, which could seal or 'lock' in the water. Bargemen ascending the canal would manoeuvre into the chamber and close the lower doors behind them. They would then lift the 'paddles' (shuttered openings) at the other end of the lock, allowing water to flow in from the higher canal. Gradually the lock filled, to the point where the water level was the same as the higher section of canal. At this point the bargemen could open the upper doors and float on their way. A bargeman descending followed the same process in reverse. Originally, this journey between Edinburgh and Glasgow meant a wearying day-long passage through 11 locks, opening and shutting 44 lock gates along the way. Hardly an easy task – and in any case the locks had since been removed. So the engineers had to do some smart thinking.

Today, if you travel west from Edinburgh along the Union Canal towards the Clyde or Glasgow, you eventually reach a place where the land drops sharply away on either side, leaving you chugging along an aqueduct that thrusts out boldly into seemingly empty space. This is the end of the Union Canal. At this point, your boat is 24 metres up in the air, roughly as high as the top of an eight-storey building. To get from this elevation down to the lower basin and float off along the Forth and Clyde

Canal, your boat must now enter the embrace of an exceptional piece of engineering, a modern take on the Celtic axe.

An immense vertical wheel (like a Ferris Wheel) 35m in diameter lies in front of your boat. The wheel has two axe-shaped arms that rotate through 180 degrees. Each arm houses a sort of 'gondola': a vessel large enough to carry two canal boats and 250,000 litres of water. A hydraulic steel gate stops the water from the high-level canal pouring out. When the wheel's gondola is aligned with the end of the aqueduct, the gate at the end of the canal and the gate at the end of the gondola open, and the boat can manoeuvre straight into the gondola. The doors are resealed – and the arms start to rotate.

At a funfair, as the Ferris Wheel turns, you'll have noticed that your seat also moves, so that you remain sitting vertically. As you travel from the bottom to the top and back again, your orientation stays the same. In a similar fashion, a complex system of gears and cogs makes sure that the gondolas on the Falkirk Wheel always remain horizontal as the arms swing through the air. To complete one 180-degree turn needs little power – the same amount of electricity as boiling eight kettles of water. This is largely thanks to Archimedes and his famous principle, which states that when an object is placed in water, it displaces its own weight. If, for example, you have one boat in one gondola, but no boats in the other side, the two gondolas will still weigh the same. The boat will have displaced an amount of water from its gondola equal to its own weight. So as long as the water levels in both arms are equal, you only need minimal power to overcome inertia and start the wheel rotating, and then momentum carries the balanced arms round until they're switched off. The Falkirk Wheel brings boats from the upper basin to the lower basin (or vice versa) in just five minutes, compared to the full day required to negotiate the canal's original system of locks.

*

There are a few examples of boat lifts around the world – such as the Strépy-Thieu in Belgium, the Niederfinow Boat Lift (the oldest working boat lift in Germany), and at the Three Gorges Dam in China (now the tallest boat lift in the world, it moves boats vertically by a colossal 113m) – but there is a particular thrill to watching and travelling on the Falkirk Wheel. Perhaps this is because it taps into our childhood memories of fairgrounds. It's an example of how engineering has an aesthetic and even nostalgic side to it that plays a part in how we respond to a structure.

No. 4: The Silk Bridge

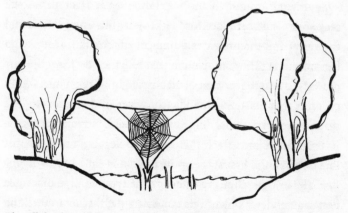

The silk bridge: the longest web-bridge in the world.

One evening I was reading a book with the television on, just letting the reassuring sound of the show's host fill the living room, but not paying any real attention. Until, that is, I heard the words 'strong material' and 'bridge' and, as you can imagine, my ears pricked up like a cat's. The host was talking about one of the most prolific bridge builders in the world – and, exceptionally, the builder is female, and lives in Madagascar.

She's about the size of a thumbnail, has eight very hairy legs and her body is heavily textured like the bark of a tree, which, as David Attenborough went on to explain, is the camouflage that protects her from predators. She also has a *spinneret*, which is the bit of her body responsible for making her the brilliant bridge engineer she is.

Darwin's bark spider can build a bridge up to 25m long (that's 1,000 times her own size), spanning rivers or even lakes. However, unlike most bridge builders, she's not looking for a way to get from one side to the other. She's looking for a way to get food.

Scurrying around in the vegetation on a river bank, she seeks out a suitable place for her project (like any professional engineer) and then releases dozens of sticky silk threads from her spinneret. They spray out – just like they do from Spider-Man's wrists in the movies – and are caught by the natural wind currents that exist above bodies of water in dense forest. The threads are carried in a thin, almost invisible, stream across the river to catch and attach themselves to vegetation there. This line of silk – the first stage of the build – is called the *bridging line*. The bridging line is a catenary, the typical curve of a cable that sags under its own weight. Giving the thread a quick tug to make sure it's secure, the spider uses the hairs on her legs, which are tiny hooks, to reel in the line a little so that it doesn't sag too much.

She tests her bridging line by walking along it, and as she does so she uses more silk and secretions to reinforce the line, making it even stronger. When she reaches the other end, she reinforces the bridging line's attachment to the vegetation by spinning more thread around it. It's important that the connection – which the wind made simply by sticking the line to a branch – will be strong enough to carry the weight of the rest of the structure.

Now the bridging line has to be anchored. The spider searches for bits of vegetation, such as large blades of grass sticking up from the water, and then moves along the line until she is almost directly above them. Slowly, producing more silk, she lowers herself and attaches her anchor point to a blade of grass close to the surface of the river, creating a 'T' shaped skeleton for her web.

Over the next few hours, the spider effortlessly shuttles back and forth, using the T-shaped skeleton as a base on which to attach more silk threads. From bridging line to anchor thread, she keeps producing and weaving new silk, in a big circular pattern. Some of the silk is not sticky: it functions as part of the structural frame of her construction. The rest is sticky, and will be the part of the web that actually traps her food. Eventually she will have fashioned a giant orb that can be more than 2m in diameter.

Darwin's bark spider is the only known spider that bridges water to trap its food. Its victims are the tasty mayflies, dragonflies and damselflies that zip over the middle of rivers. The large diameter of the web means that small creatures like birds or bats could also potentially get caught.

The sheer size of the web is hugely impressive, but the silk used to build it is even more astonishing – which makes sense: to build such a large structure you need an exceptional material. The bark spider's silk has been tested in a laboratory by connecting it to hooks that are slowly pulled apart. The results show us that these tiny creatures produce silk with an incredible *elasticity*. This is the property of a material to stretch under a load and then recover: if, after the load is removed, a material shrinks back to its original size, it is *elastically deformed*; if it doesn't fully recover to its original shape, it is *plastically deformed*. Tests have shown that the bark spider's silk is twice as elastic as other known spider silks. It is also very *tough*. Toughness is the property of a material that measures how much energy it can absorb without fracturing. It is a combination of strength (how much load the material can resist) and *ductility* (how much it can deform without breaking). In fact, the silk of Darwin's bark spider is the toughest biological material we've found so far – it's even tougher than steel.

Elasticity and toughness are a great combination for building materials. Take, for example, rubber bands: if you have a thin and stretchy band, you can stretch it very far but only with a small load, as it's elastic and ductile but not very strong. A very thick band made from brittle rubber can take more load but might snap suddenly, because it's strong but fragile. The bark spider's silk, on the other hand, has the ideal balance of all these properties. It can absorb large forces, and at the same time it can stretch a long way without snapping. This balance makes it the perfect material for building the world's largest spider webs.

*

I have included the Darwin's bark spider's bridge as a reminder that it is not just humans who create structures: in fact, as this creature demonstrates, we are still catching up with Nature. We are only now starting to build bridges that span as far as this spiders' does compared to our own body size – the Akashi Kaikyo Bridge in Japan currently holds the record for the longest span, at 1,991m. We are already being inspired by Nature (we call this type of design *biomimicry*) – the ventilation system of the Eastgate Centre in Zimbabwe is inspired by porous termite mounds, and the Quadracci Pavilion of the Milwaukee Art Museum has a retractable shade inspired by the wings of birds. But I believe we can learn even more. It would be the dream of any engineer to develop a super-material like spider silk that is incredibly tough and light, and which could be launched across a river or a valley, allowing the air to carry its threads to the other side. Then we too could create a long bridge in a few hours – just as quickly as the Darwin's bark spider does.

No. 5: Ishibune Bridge

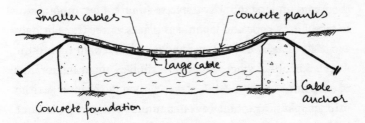

The Ishibune Bridge: a catenary bridge.

At our hotel in Tokyo, my mum and I had been given a piece of paper on which an address had been written in a series of delicate, swirling strokes that looked like little pictures. The writing was beautiful, but illegible to us, so we simply handed the paper to our taxi driver, and hoped it was enough to get us to our destination.

It was raining so hard we could barely see where we were going, but we were aware that we had left the city and were now surrounded by steep slopes covered in dense green forest. Driving higher and higher up a narrow, winding road, we finally reached a red gate with more beautiful characters inscribed upon it. Our driver came to a halt and waved us out of the car – I hoped he would still be there when we came back. I zipped up my jacket and walked along a narrow dirt path looking for the Ishibune Bridge, the perfect example of a simple stress-ribbon bridge – a form that, until I'd planned that particular trip, was unknown to me.

Earlier in the year I had been awarded a travel grant by the Institution of Structural Engineers – my proposal had been to study a special type of bridge. Speaking to my colleagues and doing some research made me aware of the stress-ribbon bridge, a graceful, simple form of which there are less than a handful of examples in the UK. I wanted to learn more about them and understand why they are so rare. My proposal was to travel to Europe and Japan – to places where these bridges are used to great effect – and report back. I went first to the Czech Republic, where engineers showed me a huge range of structures that use the stress-ribbon technique – from bridges spanning motorways to a tunnel built using the same principles. Then, at a German university, I met researchers who had built a 13m-long prototype in a lab, and who were doing tests and experiments on it. I got to do some 'testing' myself – basically jumping up and down on its deck to try to make it resonate.

To make your own mini version of a stress-ribbon bridge, use two tins of baked beans placed a metre apart to simulate bridge abutments, then lay two thick pieces of string over the tins, taping the ends to the table, which represents the ground. To turn this into a stress-ribbon bridge, it needs a deck, which you can make with matchboxes. Poke two holes through the boxes – one in each side – then lay them out on top of the strings. Thread cut pieces of rubber bands through the holes to link the matchboxes. The rubber band will stretch, compressing the matchboxes together.

If you press down on the model bridge in the middle of its span, you'll see the supporting strings tighten (in other words, develop tension); the strings pull at the tape which secures their ends to the table. A stress-ribbon bridge works in a similar way. Steel cables are slung across the gap to be bridged. The cables are thick – with a diameter about the size of a fist – and consist of numerous thin steel wires spun together to form a strong rope, which is protected by a rubber sheath. Concrete abutments at either end support the cables, which are anchored tightly into the ground. The anchors are strong enough to take the force of the cables pulling on them even when the bridge is loaded with lots of people. Planks of concrete (equivalent to the matchboxes), with grooves on the underside, are placed on top of the steel cables and connected to them to keep them in place. The planks have holes running through them, through which smaller steel cables are threaded and tightened to tie the planks together and make the deck stiffer.

The shape of these bridges reminds me of the basic rope bridges made by our ancient ancestors. Like them, and the bark spider's bridging line, a stress-ribbon bridge is a catenary. A stress-ribbon bridge is also very light – the concrete planks are quite thin at about 200 mm – and the natural curved shape of the steel cables gives them a slender and satisfying aesthetic. And, just as importantly, as far as I'm concerned, these bridges are practical too, being relatively quick to build. Once the foundations are done, the lifting of the pre-made concrete planks onto the cables is a straightforward and speedy procedure, so building them has less of an impact on the surrounding environment.

The curved red ribbon of my Japanese bridge crossed a deep ravine with a small but rapidly flowing stream at its base. As the rain hammered down, I stepped out onto the deck. It was a little bouncy. I walked up and down several times, varying my speed, then jumped on it, to see what that felt like. The movement was disconcerting, and I realised why – even though stress-ribbon bridges look fantastic and are quick to build – some people don't like them.

Because they are light and rest on cables, there's a large sag in the middle, and the ends of the bridge have relatively steep slopes which can be tricky for people with buggies or those using wheelchairs. And these bridges can be lively – their lightness and flexibility mean that, as you walk across them, they can feel unstable. Even though they are perfectly safe, stress-ribbon bridges usually move. The sag, combined with the bounciness, can give the impression that these bridges are a little dodgy. People in the three countries I'd already visited loved them, but they were used to their movement. Elsewhere, a misplaced perception of instability, and a lack of strong ground in which to anchor the tendons and keep the structure stable, might be reasons why stress-ribbon bridges haven't caught on.

By now I was soaked to the skin, but I spent a long time examining the expert engineering (after all, I had travelled nearly 10,000km to see this bridge that was so unusual back home in Britain). When the bridge shook, I clung to a handrail with one hand, trying to keep hold of my umbrella with the other. I found it difficult to stand in the middle for too long, as the valley's depth, the rapid water gushing through it, and the fact that it moved the most here, unnerved even me.

Nevertheless, like any self-respecting engineer, I made sure that my mum got plenty of snaps of me *in situ* on the structure, before we raced back to the taxi, where the driver lay asleep in his reclined seat. We drove back to Tokyo, still a little soggy from our visit.

The stress-ribbon structures I studied during my travels have stayed with me: I'm inspired by the fact that the simple rope bridge has evolved to incorporate modern technology and materials – and that despite its modernness, this new form has retained the simplicity and elegance of its forebear. New engineering doesn't always have to be big and bold; sometimes it can draw on humbler roots.

*

Bridges are all very well, but no doubt you're wondering how things worked out between me and Flirtman. All I can say is that I came to regret emailing my friend to boast that I had got rid of him in three minutes. Four years later, she read that email out loud in front of hundreds of people. During her bridesmaid's speech. At my wedding.

Yes, dear reader, I married him.

DREAM

Imagine, for a moment, a world without engineers. Abandon Archimedes. Banish Brunelleschi, Bessemer, Brunel and Bazalgette. Forget Fazlur Khan, oust Otis and, yes, get rid of Emily Roebling and Roma Agrawal. What do you see?

More or less nothing.

Of course, there'd be no skyscrapers, no steel, no elevators, no houses and no sewers under London (gross). No Shard. There'd be no phones, no internet and no TV. No cars, nor even carts – which is perhaps just as well since there'd be no roads or bridges either. And we'd be wearing more or less nothing too: there'd be no stitching together of animal skins to make clothes. And no tools for foraging, no fire for safety, no mud huts or log cabins.

Engineering is a big part of what makes us human. Sure, there are crows that can fashion a piece of wire into a hook to retrieve food, and octopuses that carry coconut shells for protection, but – so far at least – we have the edge. Engineering furnished us first with the essentials – food, water, shelter, clothing – and then with the means to cultivate crops, build civilisations and fly to the Moon. Tens of thousands of years of innovation have brought us to where we are today. Human ingenuity is bound-less; we will always aspire to manufacture more, to live better, to solve the next problem – and then the next. Engineering has created, in the most literal way, the fabric of our lives; it has shaped the spaces in which we live, work and exist.

And it'll shape our future, too. Already, I can see certain trends in engineering – irregular geometry, technology such as robotics and 3D printing, a quest for more sustainability, the merging of different disciplines (such as in biomedical engineering), a mimicking of Nature – that will once again change the way our landscape looks and feels, and the ways in which we inhabit the planet. Even if some of these seem the stuff of science fiction at the moment.

Computing capacity has made it possible for us to draw complicated, cambering shapes, such as the flowing surfaces of the Spanish Pavilion at 2010's World Expo, the undulating Guggenheim Museum in Bilbao, and the Heydar Aliyev Center in Azerbaijan, which is as intricately shaped as a conch shell. This move towards the geometrically complex takes us away from the traditional square or rectangular building and towards more natural forms. At present, creating these shapes is still expensive because it involves curving steel and shaping it into bespoke contours, or building intricate moulds for concrete. Interestingly, these moulds can account for up to 60 per cent of the total construction budget of a project – only to be binned once the concrete has hardened. In fact, to date, keeping the cost of the moulds (or *formwork*) down is one of the reasons why our columns, walls and beams tend to be rectangular: it's cheap and easy to buy rectilinear pieces of plywood.

So with this emergence of curvy shapes, we need to think smartly about how we're going to build them. (Concrete is a good option since its liquid origin makes it ideal for transforming into any shape.) One way is to use large polystyrene blocks, painstakingly carved by hand or by machine, against which concrete can be poured. But this creates much waste, because the blocks are useless once the concrete has hardened. An exciting idea – which has been around since the 1950s but has so far only been used sparingly – is the flexible membrane mould. Almost any material, ranging from hessian or burlap to light sheets of plastic made from polyethylene (PE) or polypropylene (PP), can be used. These fabrics start off slack and shapeless – but introduce some wet concrete and we're quickly reminded of what a responsive and sensory material it is: concrete interacts with the fabric, stretching and moving it to create a final shape. Two seemingly disparate materials come together in a symbiotic relationship of pressure and restraint.

Spanish architect Miguel Fisac designed the MUPAG Rehabilitation Center in Madrid (opened in 1969), using this technique to create a façade that looks cushion-like and bouncy. At one of the entrances to the Heartlands Project in Cornwall is a wall that looks like a flowing piece of silk suspended from the sky; touch it, however, and you feel solid concrete. I'm sure we're going to see more structures like this, including many on a much larger scale, because using PE or PP sheets as formwork eliminates a huge amount of waste; plus they don't tear easily – and if they do, the tears don't propagate. Moreover, nothing – including concrete – sticks to them, so they can be used multiple times. The internal steel reinforcement skeleton doesn't need to change much; neither does the concrete mix itself. But the challenge so far has been that we're simply not used to working this way. It completely changes the aesthetics of structures: architects and engineers need to catch up, as do the logistics and procurement of construction. But they will, and I bet you that when they do, I won't be the only one caught stroking concrete in public.

Talking about stroking materials: at the University of California, Berkeley, I once got my hands on some 3D printed modules (which ranged in size from my palm to a dinner plate) that could be assembled to make small installations, walls, facades and shelters. The modules were in a range of colours, and when I asked why I was gobsmacked by the answer. The white ones were salt. The black ones, recycled rubber tyres. The brown and the grey ones were more familiar materials – clay and concrete, respectively – but the purple ones were made of grape skins. That's right: grape skins. A research team led by Ronald Rael is investigating the use of unusual materials (mixed with resins to create a printable paste) to build stuff. I love the fact that, as well as working with traditional materials in a futuristic way – from geometric concrete blocks with irregular perforations to small gorgeously patterned hexagonal clay tiles for use on facades – they are also experimenting with waste materials, including those from the local wine industry. Some of their designs are self-supporting and don't require any additional structure. It got me to thinking about how 3D printing, along with exciting new combinations of materials, could lead to a future where we print these pieces and then assemble our own homes.

And 3D printing is not only being used on a modular scale – in fact, the world's first 3D-printed footbridge was opened in

Madrid in December 2016. At 12m long, it was analysed to find out exactly where the forces were being channelled; material was then deposited only in those sections – meaning minimal material, less waste and a lighter end product. Robots are also being designed to lay bricks and pour concrete on site: manufacturing embraced this trend decades ago, and now its the turn of the construction industry to catch up.

Taking the return to nature in form and material another step further is *biomimicry*, whereby not only do you mimic the *shape* of beehives, bamboo or termite mounds, but also their *function*. A famous example of this technique is the burdock burr that inspired Velcro: we copied its hooks, and its ability to stick to other surfaces. Nature builds simply and with as little material as possible, and we can reflect this principle in our structures. The skulls of birds, for example, have two layers of bone between which is a complex web of truss-like connections separated by large air pockets – in fact, bone tissue forms naturally around the cells that experience high pressure, leaving voids elsewhere. London-based architect Andres Harris conceptualised a curving canopy using a web of cushions around which a structure, similar to the birds' skulls, can be cast. Similarly, the Landesgartenschau Exhibition Hall in Stuttgart gets its inspiration from the sea urchin, which has a skeleton made from interlocking plates or *ossicles*, each of which is sponge-like and lightweight. The exhibition centre is made from 50mm-thick plywood sheets, analysed carefully by software and then fabricated robotically and assembled. If you magically expanded an egg to be the same size as this structure, the plywood would be thinner than the egg's shell.

Nature also heals itself: the human body can detect when something is wrong (often making us feel pain) and then, through a series of steps, fixes the problem. So far, with structures, we have had to intervene and perform repairs – or surgery – when things go amiss. However, a team led by Phil Purnell of the University of Leeds is designing robots that can travel – like white blood cells – through pipes in the road to diagnose defects which can then be fixed before they lead to erosion and collapse. The Institute of Making's Mark Miodownik is leading a team developing 3D printing technology that can be carried by drones to repair potholes and other road defects so we won't need to close down roads to repair them, saving money and easing traffic – the end of roadworks, perhaps? And a team at the Cambridge Centre for Smart Infrastructure and Construction is looking at adding nervous systems to new structures. A single fibre optic cable, tens of kilometres long, with continuous sensing elements, can measure the strain and temperature inside piles, tunnels, walls, slopes and bridges. Data that has never been available before can be collected, and will not only help engineers learn from these designs, but also warn them of ·impending problems.

If I try and imagine what the world of the future will look like, I imagine these biological forms interspersed with pencil-thin towers and conserved historical structures. Already, towers such as 432 Park Avenue in Manhattan boast of their slenderness ratio (it's 14 times taller than it is wide). A challenge for stability and sway, these ultra-thin skyscrapers usually have dampers. I expect we will see more and more such structures combining offices, apartments, shops and public areas as the battle for space in congested cities intensifies. Many of our historical structures are starting to underperform as time goes on: their water and drainage pipes are often inadequate; lots of heat is lost as they were never well insulated; and beams and floors can be seen to sag. Walk around London and you will notice ornate old facades shooting up into the sky seemingly unaided because the buildings behind them have been demolished. But these facades are being surreptitiously supported by a network of beams and columns that hold them steady until a new building is put in place. Using technology such as lasers to create detailed 3D maps will make it easier for engineers to understand the old and combine it with the new.

And if I really think into the future far beyond my lifetime, I imagine my descendants living underwater in pods made from paper-thin glass that cannot be shattered. Our bridges could span ten times the distance they do today because they'll be made from graphene, the 'super-material' of our future. Perhaps we will even 'grow' our homes from biological material that can be shaped and modified to suit our needs.

But for now, I arrive home every night to the welcoming arms of my old, rectangular, solid-brick Victorian flat. As I turn out the lights (still holding my rather more haggard stuffed-toy cat from New York) and begin to doze, I wonder what the Vitruvius and the Emily Roebling of the future will create. The possibilities are limited only by our imaginations – for whatever we can dream up, engineers can make real.

ACKNOWLEDGEMENTS

Thank you:

Steph Ebdon, who planted the seed of writing a book in my mind, even though I laughed and said it would never happen. I'm ecstatic that it did.

Patrick Walsh, agent extraordinaire, who believed in me and my idea, taught me how to add texture to text, and supported me through every step of the process. Leo Hollis for his support, and that timely introduction to Patrick.

Natalie Bellos, brilliant editor, who saw something in my proposal and guided me through its years of development. Her insights, dedication (even while on leave) and attention to detail are unparalleled. Lisa Pendreigh and Lena Hall for turning it into a real object, for getting it over the finish line. Pascal Cariss for making my 'sentences sizzle' – you breathed life into my words. Ben Sumner for his impeccable copy-editing. The global Bloomsbury team, for nurturing my baby and making it the book it is today.

The brilliant Mexican engineers I met: Dr Efraín Ovando-Shelley (Instituto de Ingenieria, UNAM), who showed me the Metropolitan Cathedral; and Dr Edgar Tapia-Hernández, Dr Luciano Fernández-Sola, Dr Tiziano Perea and Dr Hugón Juárez-García (Universidad Autónoma Metropolitana –

Azcapotzalco), who explained the challenges of the ground and earthquakes. The British Council, for organising a very memorable trip to Mexico City.

Phil Stride of Tideway; Karl Ratzko, Neil Poulton, Simon Driscoll of WSP; Ronald Rael of the University of California, Berkeley: for their time and interviews aiding my research. Robert Hulse at the Brunel Museum for his insight.

Rob Thomas, fountain of book-related knowledge at the Institution of Structural Engineers' library, who found the most obscure sources for me and was always available to listen to my ramblings. Debra Francis of the Institution of Civil Engineers' library for her help.

Mark Miodownik, whose book *Stuff Matters* is my inspiration (it's still on my bed-stand): the nicest guy you could ever meet, you've done so much for me. Timandra Harkness for her support, and her introduction to my wonderful writing friends at NeuWrite, who commented and critiqued.

John Parker, Dean Ricks, Ron Slade; everyone on 'The Shard Team'; the directors at WSP: for an amazing decade of learning and growing. David Holmes and Gordon Kew at Interserve; John Priestland, Mike Burton, Peter Sutcliffe and Darran Leaver at AECOM: all my superbly supportive employers – I know I am an 'unusual' employee.

David Maundrill, Joe Harris, May Chiu, Dr Christina Burr, James Dickson, Pooja Agrawal, Niri Arambepola, Emma Bowes, Chris Gosden, Jeremy Parker, Karl Ratzko and Chris Christofi, dear friends and colleagues (and sister) who read chapters, fact-checked and helped me.

The engineers and scientists, organisations and institutions out there that have inspired me to go out and tell people about what we do. For giving me a platform to speak and write. I am optimistic about the future of our collective profession – its innovation, its impact, its inclusivity.

My family, the whole extended worldwide clan – my grand-parents, aunts, uncles, cousins, nieces and nephews, and my mother-in-law – who have always been my cheerleaders, and who waited patiently for this huge project to be finished. My friends, whom I haven't seen much of recently – I will be back. You were always there. My loved ones who are no longer here, I miss you.

My parents, Hem and Lynette Agrawal; my sister, Pooja Agrawal: where do I start? For always telling me I could achieve everything I wanted through 'hard work', for inspiring me with Lego, science and our worldwide travels, for giving me the best education, for challenging and questioning me, for all your love.

And finally, my *Flirtman*, aka Badri Wadawadigi, who has navigated me – sometimes kicking and screaming – through four years of writing, for reading the words more times than anyone else, for reminding me that I can do it when I didn't believe I could, for telling me off when I procrastinated, for intelligent feedback, for naming the book, for encouraging and pushing me to achieve more than my dreams, and for your love. Let there be many more Bridges of the Day.

Roma Agrawal, February 2018

SOURCES

Addis, Bill. *Building: 3000 Years of Design Engineering and Construction*. University of Michigan: Phaidon, 2007

Agrawal, Roma. 'Pai Lin Li Travel Award 2008 – Stress Ribbon Bridges'. *The Structural Engineer*, Volume 87, 2009

Agrawal, R., Parker, J. and Slade, R. 'The Shard at London Bridge'. *The Structural Engineer*, Volume 92, Issue 7, 2014

Ahmed, Arshad and Sturges, John. *Materials Science in Construction: An Introduction*. Routledge, 2014

Allwood, Julian M. and Cullen, Jonathan M. *Sustainable Materials – Without the Hot Air: Making Buildings, Vehicles and Products Efficiently and with Less New Material*. UIT Cambridge, 2015.

Balasubramaniam, R. 'On the corrosion resistance of the Delhi iron pillar.' *Corrosion Science*, Volume 42, Issue 12, 2000

Bagust, Harold. *The Greater Genius? A Biography of Marc Isambard Brunel*. The University of Michigan: Ian Allan, 2006.

Ballinger, George. 'The Falkirk Wheel: from concept to reality'. *The Structural Engineer*, Volume 81, Issue 4, 2003

Barton, Nicholas and Stephen Myers. *The Lost Rivers of London: Their effects upon London and Londoners, and those of London and Londoners upon them*. Historical Publications, Limited, 2016

British Constructional Steelworks Association. *A Century of Steel Construction 1906–2006*. British Constructional Steelworks Association, 2006.

Blockley, David. *Bridges: The Science and Art of the World's Most Inspiring Structures*. Oxford: Oxford University Press, 2010

Brady, Sean. 'The Quebec Bridge collapse: a preventable failure.' *The Structural Engineer*, 92 (12), 2014 (2 parts)

Brown, David J. *Bridges: Three Thousand Years of Defying Nature.* London: Mitchell Beazley, 1993

Bryan, Tim. *Brunel: The Great Engineer.* Ian Allan, 1999

Clayton, Antony. *Subterranean City: Beneath the Streets of London.* London: Historical Publications, 2010.

Cross-Rudkin, P. S. M., Chrimes, M. M. and Bailey, M. R. *Biographical Dictionary of Civil Engineers in Great Britain and Ireland, Volume 2: 1830–1890*

Crow, James Mitchell. 'The concrete conundrum.' *Chemistry World*, 2008

Davidson, D. 'The Structural Aspects of the Great Pyramid.' *The Structural Engineer*, Volume 7, Issue 7, 1929. (Paper read before the Yorkshire Branch at Leeds on 7 February 1929.)

Dillon, Patrick (writer) and Biesty, Stephen (illustrator). *The Story of Buildings: From the Pyramids to the Sydney Opera House and Beyond.* Candlewick Press, 2014.

European Council of Civil Engineers. *Footbridges – Small is beautiful.* European Council of Civil Engineers, 2014.

Fabre, Guilhem, Fiches, Jean-Luc, Leveau, Philippe, and Paillet, Jean-Louis. *The Pont Du Gard: Water and the Roman Town.* Presses du CNRS, 1992.

Fahlbusch, H. 'Early dams.' *Proceedings of the Institution of Civil Engineers - Engineering History and Heritage*, Volume 162, Issue 1, 1 Feb 2009 (19–28)

'The Falkirk Wheel: a rotating boatlift.' *The Structural Engineer*, 2 January 2002

Fitchen, John. *Building Construction Before Mechanization.* MIT Press, 1989.

Giovanni, Pier and d'Ambrosio, Antonio. *Pompeii: Guide to the Site.* Electa Napoli, 2002

Gordon, J.E. *Structures: or why things don't fall down.* Da Capo Press, 2009.

Gordon, J.E. *The New Science of Strong Materials: or why you don't fall through the floor.* United States of America: Penguin Books, 1991

Graf, Bernhard. *Bridges That Changed the World*. Prestel, 2005.

Hanley, Susan B. 'Urban Sanitation in Preindustrial Japan.' *The Journal of Interdisciplinary History*, Volume 18, No. 1, 1987

Hibbert, Christopher, Keay, John, Keay, Julia and Weinreb, Ben. *The London Encyclopaedia*. Pan Macmillan, 2011.

Holland, Tom. *Rubicon: The Triumph and Tragedy of the Roman Republic*. Hachette UK, 2011.

Home, Gordon. *Old London Bridge*. Indiana University: John Lane, 1931.

Khan, Yasmin Sabina. *Engineering Architecture: The Vision of Fazlur R. Khan*. W.W. Norton, 2004.

Lampe, David. *The Tunnel*. Harrap, 1963.

Landels, J.G. *Engineering in the Ancient World*. Berkeley and Los Angeles: University of California Press, 1978.

Landau, Sarah Bradford and Condit, Carl W. *Rise of the New York Skyscraper 1865–1913*. New Haven and London: Yale University Press, 1999.

Lepik, Andres. *Skyscrapers*. Prestel, 2008.

Levy, Matthys and Salvadori, Mario. *Why Buildings Fall Down: How Structures Fail*. United States of America: W.W. Norton, 2002.

Mathewson, Andrew, Laval, Derek, Elton, Julia, Kentley, Eric and Hulse, Robert. *The Brunels' Tunnel*. ICE Publishing, 2006.

Mays, Larry, Antoniou, George P. and Angelakis, N. 'History of Water Cisterns: Legacies and Lessons.' *Water*. 5. 1916-1940. 10.3390/w5041916.

McCullough, David. *The Great Bridge: The Epic Story of the Building of the Brooklyn Bridge*. Simon & Schuster, 1983

Mehrotra, Anjali and Glisic, Branko. *Deconstructing the Dome: A Structural Analysis of the Taj Mahal*. Journal of the International Association for Shell and Spatial Structures, 2015.

Miodownik, Mark. *Stuff Matters: Exploring the Marvellous Materials That Shape Our Man-Made World*. Penguin UK, 2013.

Oxman, Rivka and Oxman, Robert (guest-edited by). *The New Structuralism. Design, engineering and architectural technologies*. Wiley, 2010.

Pannell, J.P.M. *An Illustrated History of Civil Engineering*. Univerity of California: Thames and Hudson, 1964.

Pawlyn, Michael. *Biomimicry in Architecture*. RIBA Publishing, 2016.

Pearson, Cynthia and Delatte, Norbert. *Collapse of the Quebec Bridge, 1907*. Cleveland State University, 2006

Petrash, Antonia. *More than Petticoats: Remarkable New York Women*. Globe Pequot Press, 2001

Poulos, Harry G. and Bunce, Grahame. *Foundation Design for the Burj Dubai – The World's Tallest Building*. Case Histories in Geotechnical Engineering, Arlington, VA, August 2008.

Randall, Frank A. *History of the Development of Building Construction in Chicago Safety in tall buildings*. Institution of Structural Engineers working group publication, 2002

Salvadori, Mario. *Why Buildings Stand Up*. United States of America: W.W. Norton and Company, 2002.

Santoyo-Villa, Enrique and Ovando-Shelley, Efrain. *Mexico City's Cathedral and Sagrario Church, Geometrical Correction and Soil Hardening 1989–2002– Six Years After*.

Saunders, Andrew. *Fortress Britain: Artillery Fortification in the British Isles and Ireland*. Beaufort, 1989.

Scarre, Chris (editor). *The Seventy Wonders of the Ancient World: The Great Monuments and How They Were Built*. Thames & Hudson, 1999.

Shirley-Smith, H. *The World's Greatest Bridges*. Institution of Civil Engineers Proceedings, Volume 39, 1968.

Smith, Denis. 'Sir Joseph William Bazalgette (1819-1891); Engineer to the Metropolitan Board of Works.' *Transactions of the Newcomen Society*, Vol.58, Iss. 1, 1986

Smith, Denis (editor). 'Water-Supply and Public Health Engineering', *Studies in the History of Civil Engineering*, Volume 5

Sprague de Camp, L. *The Ancient Engineers*. Dorset, 1990.

Soil Survey, Tompkins County, New York, Series 1961 No.25. United States Department of Agriculture, 1965.

Trout, Edwin A.R.. 'Historical background: Notes on the Development of Cement and Concrete', September 2013

Tudsbery, J.H.T. (editor). *Minutes of Proceedings of the Institution of Civil Engineers*

Sources

Vitruvius. *The Ten Books on Architecture* (translated by Morgan, Morris Hicky). Harvard University Press, 1914.

Walsh, Ian D. (editor). *ICE Manual of Highway Design and Management.* ICE Publ., 2011.

Weigold, Marilyn E. *Silent Builder: Emily Warren Roebling and the Brooklyn Bridge.* Associated Faculty Press, 1984

Wells, Matthew. *Engineers: A History of Engineering and Structural Design.* Routledge, 2010.

Wells, Matthew. *Skyscrapers: Structure and Design. Laurence King Publishing,* 2005.

West, Mark. *The Fabric Formwork Book: Methods for Building New Architectural and Structural Forms in Concrete.* Routledge, 2016.

Wood, Alan Muir. *Civil Engineering in Context.* Thomas Telford, 2004.

Wymer, Norman. *Great Inventors (Lives of great men and women, series III).* Oxford University Press, 1964.

https://www.tideway.london

http://puretecwater.com/reverse-osmosis/what-is-reverse-osmosis

http://www.twdb.texas.gov/publications/reports/numbered_reports/doc/r363/c6.pdf

http://mappinglondon.co.uk/2014/londons-other-underground-network/

http://www.pub.gov.sg/about/historyfuture/Pages/HistoryHome.aspx

http://www.clc.gov.sg/Publications/urbansolutions.htm

http://www.thameswater.co.uk/

http://www.bssa.org.uk/about_stainless_steel.php?id=31

https://www.newscientist.com/article/dn19386-for-self-healing-concrete-just-add-bacteria-and-food/

http://www.thecanadianencyclopedia.ca/en/article/quebec-bridge-disaster-feature/

http://www.documents.dgs.ca.gov/dgs/pio/facts/LA workshop/climate.pdf

http://www.cement.org/

http://www.unmuseum.org/pharos.htm

http://www.otisworldwide.com/pdf/AboutElevators.pdf

http://www.waterhistory.org/histories/qanats/qanats.pdf

http://users.bart.nl/~leenders/txt/qanats.html

Sources

http://water.usgs.gov/edu/earthwherewater.html

http://www.worldstandards.eu/cars/driving-on-the-left/

http://journals.plos.org/plosone/article?id=10.1371/journal.pone.0026847

https://www.youtube.com/watch?v=gSwvH6YhqIM

http://www.livescience.com/8686-itsy-bitsy-spider-web-10-times-stronger-kevlar.html

http://linkis.com/www.catf.us/resource/flbGp

http://www.bbc.co.uk/news/magazine-33962178

http://www.romanroads.org/

http://www.idrillplus.co.uk/CSS ROADMATERIALSCONTAINI NGTAR 171208.pdf

http://www.groundwateruk.org/Rising_Groundwater_in_Central_London.aspx

http://indiatoday.intoday.in/story/1993-bombay-serial-blasts-terror-attack-rocks-india-financial-capital-over-300-dead/1/301901.html

http://www.nytimes.com/1993/03/13/world/200-killed-as-bombings-sweep-bombay.html?pagewanted=all

http://www.bbc.co.uk/earth/story/20150913-nine-incredible-buildings-inspired-by-nature

http://www.thinkdefence.co.uk/2011/12/uk-military-bridging-floating-equipment/

http://www.meadinfo.org/2015/08/s355-steel-properties.html?m=1

http://www.fabwiki.fabric-formedconcrete.com/

http://www-smartinfrastructure.eng.cam.ac.uk/what-we-do-and-why/focus-areas/sensors-data-collection/projects-and-deployments-case-studies/fibre-optic-strain-sensors

http://www.instituteofmaking.org.uk/research/self-healing- cities

PHOTOGRAPHY CREDITS

P.8 courtesy of Martin Avery; P.9 courtesy of John Parker and Roma Agrawal; P.12 courtesy of Major Matthews Collection; P.18 © kokkai; P.27 © Dennis K. Johnson; P.32 © Prisma by Dukas Presseagentur GmbH / Alamy Stock Photo; P.33 © Fernand Ivaldi; P.16 © Craig Ferguson; P.38 © robertharding / Alamy Stock Photo; P.49 © Evening Standard / Stringer; P.67 © paula sierra; P.68 © Darren Robb; P.74 © Anders Blomqvist; P.80 © duncan1890; P.85 © Henry Ausloos; P.89 courtesy of Roma Agrawal; P.94 © DNY59; P.112 © Allan Baxter; P.123 courtesy of Roma Agrawal; P.133 courtesy of wikipedia; P.136 © Alvin Ing, Light and Motion; P.139 © Garden Photo World / Suzette Barnett; P.152 © Paola Cravino Photography; P.166 © De Agostini / L. Romano; P.174 courtesy of wikipedia; P.189 © exaklaus-photos; P.208 © Heritage Images; P.212 © Heritage Images; P.222 © Everett Collection Historical / Alamy Stock Photo; P.227 © Fotosearch / Stringer; P.236 © Stock Montage; P.238 © Washington Imaging / Alamy Stock Photo; P.243 © Popperfoto; P.248 © North Wind Picture Archives / Alamy Stock Photo; P.251 © Empato

INDEX

Index

Index

Index

Index

Index

Index

Roma Agrawal is a structural engineer who builds big. From footbridges and sculptures, to train stations and skyscrapers – including The Shard – she has left an indelible mark on London's landscape.

Roma is a tireless promoter of engineering and technical careers to young people, particularly under-represented groups such as women. She has advised policymakers and governments on science education, and has given talks to thousands worldwide at universities, schools and organisations, including two for TEDx. She is a television presenter, and writes articles about engineering, education and leadership.

Roma has been awarded international awards for her technical prowess and success in promoting the profession of engineering, including the prestigious Royal Academy of Engineering's Rooke Award.

Built is her first book.

www.RomaTheEngineer.com
@RomaTheEngineer

A NOTE ON THE TYPE

The text of this book is set in Minion, a digital typeface designed by Robert Slimbach in 1990 for Adobe Systems. The name comes from the traditional naming system for type sizes, in which minion is between nonpareil and brevier. It is inspired by late Renaissance-era type.